Geographic Infomation Systems

G I S
Fundamentals, Applications and Implementations

K Elangovan
Sr. Lecturer, Deptt. of Civil Engineering
PSG College of Technology
Coimbatore

New India Publishing Agency
Pitam Pura, New Delhi- 110 088

© Publisher, 2006

All rights reserved, no part of this publication may be reproduced, stored in a retrieval system or transmitted in any form or by any means, electronic, mechanical, photocopying, recording or otherwise without the prior written permission of the publisher.

ISBN 978-81-89422-16-5

Published by:

New India Publishing Agency

101, Vikas Surya Plaza, CU Block, L.S.C. Mkt., Pitam Pura, New Delhi- 110 088, (INDIA)
Phone: 011-27341717, Fax: 011-27341616
E-mail: spjain_niph@rediffmail.com
Web: www.bookfactoryindia.com

Dedicated

to my

Parents

Krishnan and Janaki

PREFACE

Geographic Information System is the new area with diversified applications in Civil Engineering, Geosciences, Forestry, Disaster mitigation, Environment and Ecology, Infrastructure planning, Utility mapping, Business, Mobile mapping, Information Technology and in many more fields. Since nearly 80% of the real world data are spatial in nature, GIS technology has been popularized overwhelmingly. It is not overestimation that GIS will be used in day to day life within next decade. Already developed countries use GIS widely and it will be used in day to day activities in developing countries also in near future.

Reduced cost in hardware and software in GIS made it possible to reach most of the government organizations, research institutions, academic institutions and non governmental organizations. Although lot of data are available in paper format, if they are not collected, stored in a structured way for visualizing the real world, then the data do not serve any purpose. But this bottleneck has been eliminated by using GIS. Mobile mapping is the new area in which dynamic GIS maps are used in cell phones to find the actual location of a person. Anyone shall find a persons exact location when the cell phone is integrated with GPS and GIS maps. This will become reality in all parts of world in this decade.

It is important to use any tool with its background knowledge. Also GIS is a tool and it has been used as an attractive tool nowadays. It is very much important to know the basics of GIS and their applications. Hence to explain the concepts of GIS in a simple language, this book has been attempted. Thus the book covers basics of data types, database concepts, map projections, errors and removal of errors, advanced analysis, web GIS, implementing a GIS project, *viii*

various software and hardware requirement and advanced GIS applications. Hence any user in GIS will find this book as a catalyst to understand GIS. Also a GIS resource bank is provided which will give the users more information. Glossary consists of definition of important GIS terminology. It is advisable to read the glossary before actually going through the book, hence the readers will understand the GIS easily.

The aim of this book is to provide a textbook on GIS in a simple style for the beginners in diversified disciplines. This book shall be used as a text book for undergraduate, post graduate students, GIS users and researchers.

September 10, 2005 **Dr. K Elangovan**

ACKNOWLEDGEMENTS

I would like to thank Shri.V.Rajan, Managing Trustee and Shri.C.R.Swaminathan, Chief Executive, PSG Institutions for the facilities provided at the Institution.

Sincere thanks are due to Dr. R. Rudramoorthy, Principal Incharge, PSG College of Technology, for the facilities and support given to me. Also I thank Dr. J. V. Ramasamy, Prof. & Head, Department of Civil Engineering, PSG College of Technology, for his constant encouragement.

I would like to express my gratitude to Dr. P. Radhakrishnan, Former Principal, PSG College of Technology and Present Vice Chancellor, Vellore Institute of Technology, Vellore for his encouragement and being a source of constant inspiration.

I thank Dr. S. Rajasekaran, DSc, Former Head and presently Professor, Department of Civil Engineering and Dr. P. V. Mohanram, Prof. & Head, Department of Mechanical Engg., PSG College of Technology for their encouragement.

I thank all my faculty colleagues of Department of Civil Engineering, PSG College of Technology, Prof.Surendranath Muller, Dr. P. Parameswaran, Prof.R.Ragupathy, Dr. G. Sankarasubramanian, Dr. M. Palanikumar, Ms. K. Nalinaa, Ms. A. Thilagam, Mr. M. Ibrahim Bathusha, Mr. C. G. Sivakumar, Mr. P. Senthilkumar, Mr. M. Manivannan, Mr. A. Rajkumar, Mr. T. Ramesh and Mr. S.Anandakumar.

I thank Dr.L.S.Jayagopal, and Dr. S. C. Natesan Former Prof. & Heads of Department of Civil Engineering, PSG College of Technology for their encouragements.

I thank all the students of Department of Civil Engineering, PSG College of Technology for their useful interaction with me during my teaching which has refined this book.

My sincere thanks and appreciation goes to M/s. New India Publishing Agency, New Delhi, for their idea of writing this book.

I would like to express my deepest thanks to my wife Sudharitha and daughter Poojashree.

CONTENTS

Preface	*vii*

CHAPTER 1: GIS — AS A SCIENCE AND TECHNOLOGY   1

1.1   Introduction ... *1*

1.2   Geographic Information System (GIS)– Terminology ... *2*

1.3   Creation of GIS Database ... *3*

1.4   Spatial Query in GIS ... *3*

1.5   Development of GIS ... *5*

1.6   Characteristics of Maps ... *11*

1.7   Map Projection ... *18*
       1.7.1 Cylindrical Projection
       1.7.2 Universal Transverse Mercator (UM) Projections
       1.7.3 Conical Projection
       1.7.4 Azimuthal Projection

1.8   Representation of Earth Features in GIS ... *23*

1.9   Components of GIS ... *25*

1.10   Data for GIS ... *28*

CHAPTER 2: APPLICATIONS OF GIS AND RESOURCE MANAGEMENT   33

2.1   Introduction ... *33*

2.2   Civil Engineering ... *34*

2.3   Environment ... *34*

2.4   Disaster Management ... *35*

2.5   Emergency Planning ... *35*

2.6   Geology ... *36*

2.7   Utility Management ... *36*

2.8   Forestry ... *37*

2.9   Agriculture ... *37*

2.10   Business... *38*

2.11   Land Information System and Cadastral Records ... *38*

2.12   Military ... *39*

xii

CHAPTER 3: **GIS TECHNIQUES AND NATURE OF DATA**   **41**

3.1   Introduction ... *41*

3.2   Data Structures ... *44*
  3.2.1 Raster Data Structures
  3.2.2 Vector Data Structures

3.3   Raster Data Input ... *54*
  3.3.1 Remote Sensing

3.4   Vector Data Input ... *66*
  3.4.1 Global Positioning System (GPS)

CHAPTER 4: **DATABASE CREATION AND ANALYSIS IN GIS**   **77**

4.1   Introduction ... *77*

4.2   Database Concepts ... *78*
  4.2.1 Hierarchical Database
  4.2.2 Network Database
  4.2.3 Object Orientated Model
  4.2.4 Relational Database Concept
  4.2.5 Database Design

4.3   Data Analysis ... *84*

CHAPTER 5: **ERRORS IN GIS AND DATA OUTPUT**   **97**

5.1   Introduction ... *97*

5.2   Original Sources of Errors ... *97*

5.3   Errors Introduced Due to Data Processing in GIS ... *98*

5.4   Errors in Method ... *101*

5.5   Methods of Correcting Existing Errors ... *101*

5.6   Data Output ... *104*

5.7   Classification Method for Graduated Colour Map of
      Graduated Symbol Map ... *108*

CHAPTER 6: **ADVANCED GIS APPLICATIONS** — Case Studies   **111**

6.1   Introduction ... *111*

6.2   Network Analysis ... *111*

6.3   Spatial Analysis ... *114*

6.4   Terrain Analysis ... *115*

6.5   Case Studies ... *124*
  6.5.1 GIS Based Water Distribution System ... *124*

6.5.1.2 Study Area
6.5.1.3 Data Collected
6.5.1.4 Methodology
6.5.1.5 Querying the Maps

6.5.2 Route Optimization for Solid Waste Disposal ... *130*
6.5.2.1 Introduction
6.5.2.2 Study Area - Coimbatore Municipal Corporatio
6.5.2.3 Objective
6.5.2.4 Methodology

6.5.3 Site Selection for Solid Waste Disposal ... *136*
6.5.3.1 Introduction
6.5.3.2 Methodology
6.5.3.3 Landuse or land cover
6.5.3.4 Permeability of Soil

6.5.4 Groundwater Quality Assessment Using GIS ... *140*
6.5.4.1 Introduction
6.5.4.2 Study Area
6.5.4.3 Materials and Methods

6.5.5 Site Suitability Analysis Using GIS ... *143*
6.5.5.1 Introduction
6.5.5.2 Methodology
6.5.5.3 Result

6.5.6 Soil Erosion Modeling for Coimbatore District ... *151*
6.5.6.1 Introduction
6.5.6.2 Factors Affecting Soil Erosion
6.5.6.3 Methodology
6.5.6.4 Conclusion

6.5.7 Groundwater Level Variation Level Analysis ... *156*
6.5.7.1 Introduction
6.5.7.2 Methodology
6.5.7.3 Conclusion

6.5.8 Groundwater Quality Assessment Using GIS ... *157*
6.5.8.1 Introduction
6.5.8.2 Methodology
6.5.8.3 Conclusion

6.5.9 Rejuvenation of Noyyal River Using GIS ... *160*
6.5.9.1 Study Area
6.5.9.2 Data for the Study
6.5.9.3 Tributaries of River Noyyal
6.5.9.4 Anaicuts of Noyyal Basin

xiv

6.5.9.5 Methodology
6.5.9.6 Results from Erdas Imagine 8.4 Analysis
6.5.9.7 Comparative Study in Encorachments
6.5.9.8 Results from ArcView 3.2a Analysis

CHAPTER 7: INTERNET GIS 175

7.1 Introduction ... *175*

7.2 Hardware and Software ... *175*

7.3 Internet GIS Architecture ... *177*

7.4 Internet GIS Applications ... *177*

7.5 Future of Internet GIS ... *177*

CHAPTER 8: PROJECT IMPLEMENTATION 179

8.1 Introduction ... *179*
8.2 Problem Identification ... *179*
8.3 Infrastructure Requirement ... *179*
8.4 Organization and GIS Experts ... *181*
8.5 Data Source ... *185*
8.6 Questions to be asked before purchasing a GIS software ... *181*
8.7 Open GIS ... *182*
8.8 Evolution in GIS Standard ... *182*
8.9 Future of GIS ... *182*

COLOUR PLATES	185
GLOSSARY	189
GIS RESOURCE BANKS	201
BIBLIOGRAPHY	205
INDEX	209

Geographic Information Systems (GIS) — As A Science and Technology

1

1.1 INTRODUCTION

Geographic Information System (GIS) is the new emerging field and grows at very rapid pace. Now GIS is a billion dollar industry with applications in varied disciplines. At present, GIS is being used by professionals from various disciplines. In this decade, GIS is being used by experts who are good in GIS technology for various applications. The development trend indicates that GIS will be used by common man in near future and at the same time, advanced GIS analysis will be carried out by GIS experts.

Any technology is developed out of necessity and GIS is of no exception. When a large amount of data were available for water distribution systems, road network, sewer lines, telecom lines, electricity lines, gas pipe lines, state wise population, natural resources *etc*, it is important to store, maintain and retrieve the data for applications. It is not possible to manually browse through the large amount of paper records. Also even if the data is available in an unorganized way in the computerized environment, it is not useful. Hence if the data are referred to locations in earth surface and the data are stored in an organized way, it is possible to retrieve and analyse the data for applications. Hence the solution is GIS and developed today into reality. Developed countries are using GIS widely in many areas but developing countries are moving towards the development of GIS database for their resources.

Remote sensing, aerial photography, cartography, surveying and other field instruments for attribute data collection contributes to the data acquisition. Cartography, surveying, geography, geodesy contributes for mapping process. Disciplines like computer science and statistics, mathematics involved in processing and analysing data . Computer science and mathematics involved in storing the data structure.

GIS maps are intelligent. It has many advantages compared with paper map in this digital age.

Spatial data means data which are referenced to earth. Maps, satellite imageries, aerial photographs are example of spatial data. Attribute data which are attached to spatial data are called as aspatial data or non-spatial data.

If these data are brought into GIS then it is stored in a standard format, it is possible to update, share, retrieve, manipulate and analyse quickly. Much of the time and money is saved in case of reproduction and better decisions could be made

Creation of GIS database is a time consuming job. Once the GIS database is created, the results are obtained by simple mouse clicks. Once a GIS database is created it will be useful for long period and updating is easy.

Geographic Information Systems is offered as individual discipline in the name of Geomatics, Geoinformatics and Spatial Information Technology. These programmes are offered at UG and PG level. Much research activity is also being undertaken. This field has become an individual discipline now.

1.2 GEOGRAPHIC INFORMATION SYSTEMS (GIS) – TERMINOLOGY

GIS is expressed in individual letter G-I-S and not at pronunciation GIS. Also it is geographic or geographical information systems.

Geography is dealing with the real world with their data. Information is the data and its use. System is the computer with accessories with well established methodologies.

Definition

GIS is a computer based technology and methodology to collect, store, manipulate, retrieve and analyse spatial data or georeferenced data.

or

GIS is a system of hardware, software, data, people, organisation and institutional arrangement for collecting, storing, analyzing and displaying information about the areas of earth

GIS require vast amount of data. Data is converted into information. Temperature of Chennai City for a particular day is 30° and for New Delhi it is 20°. Here 20° and 30° C are data. But the information is that Chennai is hotter than New Delhi. GIS projects are costly because of the large data requirement from various sources. Decision with less information is not complete.

1.3 CREATION OF GIS DATABASE

Data for GIS come from satellite imageries, maps, GPS, survey data and other instruments for attribute data. The data are entered into computer using scanner, digitiser, floppy, Cd, internet, intranet and keyboard.

Attributed data are entered in the database management system present in GIS or in the database created with external database softwares can be linked with GIS softwares. Manipulation is changing the data from one format to other, one map projection to other and removing errors. With the available analytical functions available in GIS, the data are analysed and displayed. The data is retrieved and viewed in screen or taken as printout. These results are stored in Cd, hard disk, floppy, internet for further analysis, planning and decision making. GIS is a very good decision support system.

1.4 SPATIAL QUERY IN GIS

Once a GIS project is created, it is possible to impose different types of queries to obtain results to our expectation. Not only simple queries, many complex queries, analysis, modeling and forecasting are possible in GIS.

A tourist may find a suitable hotel within 5km radius of airport using GIS. A person who is going to reside in a new city may be interested to find a suitable house by asking questions like, find houses within 5km radius of his working place, an hospital nearby, a shopping complex nearby and a school nearby. Results will be displayed based on his query and from the result, the person may decide to select the houses which are suitable to his requirement.

The following questions may be asked and answers may be found using GIS in various fields. The questions given are simple but one may ask complex question by combining more than one question. Many more queries can be asked for varied applications.

- What is the area of the Ooty lake?
- Where is the No.A13 lamp post?
- What is the relation between slope and landslide?
- How long is the river between two locations?
- What is the shape of the landuse?
- Where from the water comes?
- How many 3 star hotels are available within 5km radius of airport?
- What is the elevation at my place?
- How much will the expected rainfall in my district tomorrow?
- How many isolated waste lands with 300 acres available for dumping the solid waste?
- Where shall we dispose the nuclear waste safely?
- What is the relation between elevation and species diversity?
- Where shall I lay a road in the hilly terrain
- Where landslide may occur?

Geographic Information System 5

- What is the erosion rate in Nilgiris?
- What is the rate of sedimentation in the reservoir?
- Where shall the high income group available for my business?
- Which is the direction of forest fire for the next hour?
- Where do I select a house for rent?
- Where shall I locate the Telecom Cable near the road for this street?
- What is the optimum route for dumping solid waste for my city?
- Where does the endangered species live?
- What exists at this location?
- Where is good quality groundwater in a district?
- What is the urban development pattern in Chennai for the past 20 years?
- What are the areas affected if a landslide occur in an area?
- What is the shortest path between one location to other location?
- What is the cost path to lay the gas pipeline from one location to an other location?
- What are the 5 star hotels within 7km radius of chennai airport?

To perform the above questions a strong GIS database is required

1.5 DEVELOPMENT OF GIS

GIS has been developed due to the combination of many disciplines. The following disciplines contributed towards the development of GIS

- Computer Mapping
- Databases
- Computer Science

6   *GIS: Fundamentals, Applications & Implementations*

— Geography
— Remote Sensing
— Data Processing
— Mathematics and Statistics
— Computer Aided Design(CAD)
— Cartography

Computer mapping is nothing but creation of map in computer and storing it in graphical form. Databases are used to store, manipulate, retrieve the data. Computer science is helpful in the development of computers in processing, storage capacity and speed. Geography is used to locate the real world features in terms of latitude, longitude altitude and projecting the features using the map projections. Remote sensing is taking satellite imageries of earth from higher altitude for computing the earth's dimension and also for using the natural resource of earth. Data processing is processing spatial and non spatial data using well established algorithms. Mathematics and Statistics are developed for analysis of numerical data. Computer Aided Design (CAD) is used to draw map in computer using the CAD tools. Cartography is used to draw map in computer using the well laid cartographic rules. GIS is developed by using all the above disciplines and emerged into a new field.

GIS was used in a few universities and organizations during 1960s. Today GIS is a component of IT and used in diversified disciplines. GIS require spatial data. Data is converted into information. Knowledge is converted into intelligence. Data is used both as singular and plural. But datum is singular and data is plural. Conventionally data is being used for both singular and plural. GIS is useful to develop spatial intelligence for decision making. GIS is a decision support system.

Romans first employed the concept of cadastral records *i.e.* land records. Egyptians also used property registration for tax revenues. Maps were used for sea voyages. Arabians were the leading cartographers during middle ages. Expeditions of a Marco Polo,/Christopher Columbus, Vasco-de-Gama and others resulted in increased trade. Due to development in European countries and new discovered regions, the need of geographic information increased. Military agencies were responsible for creating maps in many countries. Until 19th Century , maps were used for navigation, trade, taxation and military. But now-a-days maps are used in every walk of life. First geological map of Paris was compiled in 1811. Irish Government in 1838, compiled a series of maps for use of railway engineers and regarded as first manual GIS. During 1909, first aerial photographs were used for making maps. After two World wars, aerial photographs were used for making detailed maps in the scale of 1:5000 to 1:50,000.

GIS concept is being used in one or other form from a long time back. But due to the advent of computers, the term GIS was introduced and is being used widely now-a-days. During Roman Empires period, Land surveyors (Agrimensors) used maps for land administration. In 17th Century, Mercator (Surveyor) arrived calculations for map projections. In Early times maps were used for navigation, route finding and military purposes. But later maps were used in civilian applications. In 18th Century European states formed Geographic Information history through national bodies to produce cadastral and topographical maps.

Geodesy, Photogrammetry and Cartography are called as Mapping Sciences. Geodesy deals with the survey of earths surface with terrestrial survey equipments. Photogrammetry deals with deriving contours and other topographic features using aerial photographs. Cartography is the art and science of drawing maps. All the above disciplines are considered in producing a map.

In 17th Century, Copper plates were used for printing colour symbols. Qualitative developments started in early period and later quantitative developments took place. During 1930s appropriate mathematics for spatial problem developed and in 1940s Statistical and Time Series analysis came into existence.

A Manual method that combines several map layers to identify suitable sites by meeting a number of criteria called suitability. Mapping or sieve Mapping was introduced by Ian

McHarg (McHarg Overlay, 1960, McHarg-Landcape Architect). Here various layers of maps like contour, drainage, soil type, settlements, landuse/landcover maps were prepared in transparency sheets and kept over a light tracing table. Then a road was drawn which follows certain gradient, connecting many settlements, not passing through many rivers and passing through wastelands. A new road is formed in this way. Such mappings were done before GIS was introduced. Nowadays drawing such a road is done automatically within few seconds if all the layers are available in GIS.

In US, the Public Health Service, Forestry Service, Bureau of Census, Harvard Graphics Lab played crucial role in development of GIS. Many Individual researchers have carried out many work on GIS around the world. In 1960s little commercial development took place but British Oxford Experimental Cartography and some others in other parts of world used digital technology. Computer Aided Design (CAD) was developed for drawing maps and solid diagrams. Later maps in CAD were upgraded by adding some more details and querying. They are called as AM & FM (Automated Mapping and Facility Management). Such AM/FM were used for maintaining many utilities like water distribution systems, sewer systems, telecom lines. Later many more analytical functions were added in AM/FM and GIS emerged. AM & FM was behind the development of GIS. GIS attempts to produce a computer model of the real world to assist problem solving and decision making.

GIS origin dates back to 1940s and 1950s. Successful implementation of computer aided graphical data processing at Massachusetts Institute of Technology and database management system by General Electric in 1965 paved the way for GIS development at faster rate. Government agencies in USA, Canada, UK started using GIS for processing large amount of data. During 1960s, CGIS developed and regarded as first GIS used by Canada government agencies. This was used for land and resource management. USGS used Geographical Information Retrieval and Analysis System (GIRAS) to analyse landuse and landcover data. During 1970s, Swedish Land Data Bank took steps to automate the land and property registration. Harvard Graphics Laboratory developed first vector GIS called as ODESSEY. Center for Urban and Regional Analysis, University of Minnesota, developed Minnesota Land Management Information System. GIMMS (Geographic Information Mapping and Management System) developed by department of Geography, University of Edinburgh, Scotland

Computer cartography is also called as Automated cartography, Digital Mapping and Computer Aided cartography. In 1963 Atlas of Britain was printed using computer by Bickmore & his team at Oxford Experimental Cartography. During this period first free cursor digitizer, program for line measurement, projection changes and data editing available

In 1965 in US First Automated cartography package called SYMAP (Synagraphic Mapping System) made available and it can produce chloropleth maps(Map showing equal distribution of an entity). Later SYMAP was improved called SYMVU used until 1980s. Spatial statistics evolved in 1950s and 1960s for measuring spatial distribution, 3D analysis, network analysis and modeling techniques. GIS work was carried out in Canada during 1964. In 1970 US bureau of census started a good approach towards GIS. Then Harvard graphics Lab, US led to the production of Commercial GIS. 1970s - Automated cartographic package like GIMMS, MAPICS & SURFACE II, GRID, IMGRID, GEOMAP, MAP available. 1970 First multicolour map was published by Britain. Soon after Ordance survey in UK published series of multicolour maps. During 1970 Blind digitizing lost and interactive user control digitizer started. First conference on GIS and published work on GIS available during 1970s. First meeting of academics to discus about GIS held at UK during 1975 and first text on GIS was released by International Geographical Union published in the same year.

During 1970s to 1980s , hundreds of packages developed related to GIS. Earlier period, many universities, government agencies and R & D organizations used GIS. Raster GIS was developed during initial stages of GIS development. After the concepts of GIS established and also due to development in computer science lead to the development of vector GIS. During 1960s to 1970s mainly mapping packages were developed with little analytical capability. During 1979, due to the development in topology concept, many vector packages developed with inbuilt topology. This is regarded as a milestone in GIS history. In 1982, Environmental Systems Research Institute (ESRI) released its vector based GIS ArcInfo with georelational model. GIS softwares similar to this called INFOMAP developed by Synercon in US and CARIS by Universal System Ltd in Canada. Late 1980s, Packages like MapInfo, SPANS, PC ArcInfo developed. Intergraph released its Modular GIS Environment during 1989.

During mid 1970s, the developments were bifurcated such that one branch of GIS package development and other branch for developing cartographic packages.But in 1980 , merging of GIS and Cartography happened. Later high speed computers were available. Graphics and query were possible. In1986 Intel 80386 processor was available for Microcomputers and in 1990s Raster data structure & Quadtree data structure developed. In 1987 International Journal on GIS released.

The development in GIS shall be grouped into four developmental stages. During early 1960-1975, GIS concept was introduced and few people only used GIS in mainframe computers. During 1973 to early 1980s, experiment and practice was done at various parts of world and much of the works were duplicates as they were done by many individuals. But from 1982 to late 1990s, many softwares on GIS came into existence and GIS growth was good. During 1990 to 2000, GIS technology was standardized, software made available for various platforms and many are user friendly software. Unlike earlier GIS which were command driven, during this period GIS used graphical user interface for working. After 2000 GIS is used in virtual reality, fly through, multimedia integration, Mobile mapping and web based GIS. Use of expert system with artificial intelligence concept developed.

National Spatial Data Infrastructure (NSDI) introduces during 1994. Later Open GIS Consortium developed. GIS is coming with multimedia tools and virtual reality is possible in GIS at present.

1.6 CHARACTERISTICS OF MAPS

Scale, resolution, accuracy and projection are the important component of a map.

Scale

Real world features are represented in the map to scale. Graphical scale, numerical scale are available.

A map may consists of the scale of 1: 24,000. Here 1: 24,000 means 1 inch in maps equals to 24,000 inches in ground or 1 foot in map equals to 24,000 feet on ground or 1 meter in map equals to 24,000 meters in ground. No map is 100% accurate. As the maps is the abstraction of the real world features, it consists of errors. Graphical scales are also used which are useful to compare the map visually with ground (Fig. 1.1).

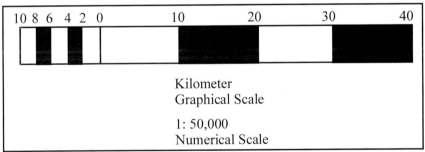

Figure 1: Types of Scale.

Small scale map is 1:1,000,000 and large scale map is 1:500. Small scale maps shows large area of earth surface with smaller details and large scale map shows smaller area with more details. (Fig 1.2). In India, Survey of India (SOI) head quarter at Dehra Dun, is responsible for the survey, production and distribution of maps. SOI publish maps in three scales viz., 1:25,000, 1:50,000 and 1: 2,50,000. They are called as survey of India toposheets. In western countries such maps are called as quadrangle sheets.

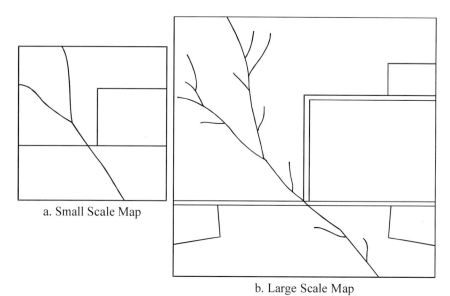

Figure 1.2: Relation between Small Scale map and Large Scale map.

These toposheets consists of very detailed information of the countries. Presently Survey of India supplies the toposheet in digital format (vector) which can be used without further digitization.

Resolution: The minimum size of an object that is represented is called as resolution or minimum mapping unit. The relation between scale and resolution is shown in the Table 1.1.

Table 1.1: Relation between scale and resolution

Scale	Resolution in m
1:500	0.01
1:1000	0.02
1:5000	1
1:10000	2
1:25000	5
1:50000	10
1:100000	20
1:2,50000	50
1:1000000	200

Map: A map is a representation of cartographic abstraction of real world features on the map to scale on a paper

The features in the real world are generalized and represented in a map with cartographic symbols. Various types of maps are available based on their usage.

Types of Map

Topographical map: A map showing the surface features of the earth's surface (contours, roads, rivers, houses etc) in great accuracy and detail relative to the map scale used.

Thematic map: A map displaying selected kinds of information relating to specific themes such as soil, landuse, population density, suitability for arable crops and so on

Geological maps: Maps showing rock types and geological structures

Relief maps: Maps showing contours

Agricultural maps: Maps showing crop distribution

Cadastral maps: Map showing land records

Navigational charts: Maps used for sea and air routes

Political maps: Maps with administrative boundaries of countries

Commercial maps: Maps showing the business details

Weather maps: Maps showing temperature, pressure, rainfall, snow fall and wind speed details.

Coordinate system

Earth is not a perfect spheroid. It is oblate spheroid with undulations. Geoid, projection, coordinate system and spheroid are the terms used in map projections to represent the features accurately in maps. Two forces are acting on earth. One is the gravity towards centre of the earth and other due to the spinning motion of earth. This is more at equator and negligible at poles.

At some surface two forces are equal. Geoid refers to the line which is close to mean sea level. Geoid is irregular surface. But spheroid is an approximation of Geoid. Many spheroids are available for countries of the world. Locations are expressed in latitude and longitude which is measured in relation to Geoid. Hence it is important to correct the difference between Geoid and Spheroid for locations. (Fig 1.3)

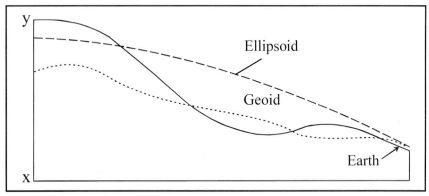

Figure 1.3: Geoid - ellipsoid relation

A reference system used to measure horizontal and vertical distances on a planimetric map. A coordinate system is usually defined by a map projection (its units and characteristics), a spheroid of reference, a datum, one or more standard parallels, a central meridian, and possible shifts in the x and y directions. It is used to locate x,y positions of point, line, and area features.

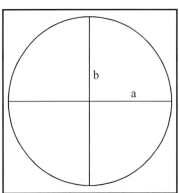

Map projections allow areas on the surface of the Earth (a spheroid) to be represented on a map (a flat surface). In this way the precise position of features on the Earth's surface can be obtained from the map.

Earth is a spheroid flattened at poles. Hence polar axis is 23 km shorter than equatorial axis (Fig 1.4)

Figure 1.4: Semi major and semi minor axis.

Polar flattening (f) = (a-b)/a
Where a is major axis and b is minor axis.
f- ellipticity = 1/298

Earth is referred with degree, minute and seconds (DMS). One degree consists of 60 minutes and 1 minute consists of 60 seconds.

In most business applications, the characteristics of the map projection being used will probably not be of critical importance. Business applications, for example, are typically concerned with the relative location of different features, such as sources of supply, demand, and competition, rather than their absolute location on the Earth. On large scale maps, such as street maps, the distortion caused by the map projection is less. On smaller scale maps, such as regional and world maps, where a small distance on the map may represent a considerable distance on the Earth, this distortion may have a bigger impact.

Knowledge of the characteristics of the map projection is more important.

To make mathematical calculations easier, the Earth is often treated as a sphere, having a radius valued at 6,370,997 meters. This assumption that the Earth is a sphere can be used for small-scale maps, those less than 1:5,000,000. At this scale, the difference between a sphere and a spheroid cannot be detected on a map; however, to maintain accuracy for larger-scale maps (scales of 1:1,000,000 or larger), the Earth must be treated as a spheroid. Spheroids are also referred to as ellipsoids.

Locations of a feature is referred in coordinate systems. Three types of coordinate system are available. They are Cartesian coordinate system, Planepolar coordinate system and Global coordinate system.

Fig 1.5 show the plane orthogonal Cartesian coordinate system. OX is the ordinate and OY is the abscissa. A point A is marked by the coordinates of X and Y. Fig 1.6 show the plane polar coordinate system. OQ is the horizontal line and with O the vectorial angle line OA is drawn for a radius r.

16   GIS: Fundamentals, Applications & Implementations

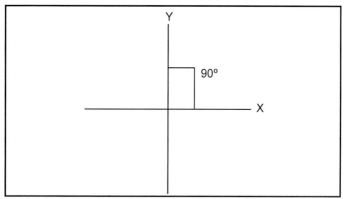

Figure 1.5: Cartesian coordinate System

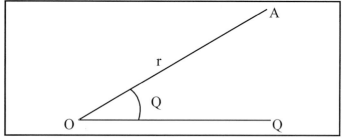

Figure 1.6: Plane polar coordinate system

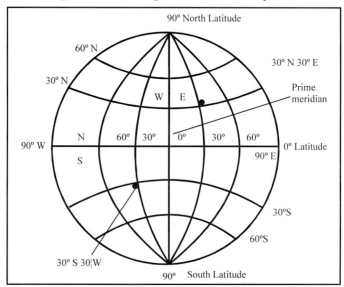

Figure 1.7: Global Coordinate System

Fig 1.7 show the global coordinate system. Lines passing through north south are called as longitudes and lines passing

Geographic Information System 17

through east west are called as latitudes (also called as parallels). Totally 360° longitudes are available. Latitudes are 90° towards north and south from equator. Hence a location in earth is represented using the latitude and longitude. 0 is Greenwich meridian. Longitude extend 0° to +180° in east and 0° to -180° in west.

Following are some of the spheroids commonly used with their description. For India Everest 1830 spheroid is used. For Russia, Krasovsky 1940 is used. For Australia, Australian spheroid is used. The spheroids are developed for particular part of the world or area of the world. It is important to use appropriate spheroids for particular areas. Otherwise the calculations of the earth surface details will be erratic.

Spheroid	Description
Airy	Airy 1830
Australian	Australian National 1965
Bessel	Bessel 1841
Clarke 1866	Alexander Ross Clarke 1866
Clarke 1880	Alexander Ross Clarke 1880
Everest	Everest 1830
GRS80	Geodetic Reference System 1980
International 1909	John Fillmore Hayford 1909 (International 1924, IUGG 1924)
Krasovsky	Krasovsky 1940
Sphere	The world as a sphere. Use this for world maps
WGS72	World Geodetic System 1972
WGS84	World Geodetic System 1984

Datum is a set of parameters defining a coordinate system, and a set of control points whose geometric relationships are known, either through measurement or calculation. One part of defining the coordinate system is the spheroid used to approximate the shape of the earth.

A spheroid is defined by a radius and an eccentricity. These two constants are used as inputs to the equations which calculate a projected coordinate from a coordinate in decimal degrees. When a projection is created, it is associated with a default spheroid so that these constants will be available. This default spheroid varies from projection to projection, but is usually the Sphere for small-scale projections and Clarke 1866 for large-scale projections.

1.7 MAP PROJECTION

Map projection is a mathematical transformation that is used to project the spherical 3 dimensional earth surface on a flat 2 dimensional surface

Three levels of map projections are considered.

1. Class
2. Aspect
3. Property

Basic class of map projection

1. Cylindrical projection
2. Azimuthal projection
3. Conical projection

Aspect of map projection

1. Normal aspect
2. Transverse aspect
3. Oblique aspect

Properties of map projection

1. Conformity - same angle on all sides
2. Equivalent - equal areas
3. Equidistance- equal distance in 1 direction

Selecting suitable map projection

Cylindrical projection is used for Countries in tropics, conical projection is used for countries in temperate regions, and azimuthal projection is used for countries in polar regions.

For preparing large scale topographic maps, Transverse Mercator projection is used. Polyconic projection and Lambert conformal conical projection are used in smaller extent.

93% of earth is mapped with the four ellipsoids namely International, Krasovsky, Bessel and Clarke (1880). Other ellipsoids used are Clarke 1866, World Geodetic System (WGS) 72, and Geodetic Reference System (GRS 80).

The following are the properties that are affected any projection.

Conformality : Scale is same in the map in any direction. Latitudes and Longitudes intersect at right angles. Scale is preserved

Distance : From the centre of projection, distance is same to other places

Direction : From one point to other point, azimuth is same

Scale: Scale is same in the map and also in the ground

Scale is the relationship between a distance portrayed on a map and the same distance on the Earth.

Area : is same in map and ground.

1.7.1 Cylindrical projections

When a bulb is inserted into a transparent 3D globe which is kept in a cylinder, when light glows, the features of the earth are displayed on the cylinder. The cylinder may be cut vertically and opened. This consists of all the features of earth. When the equator is touching the cylinder, it is tangent case. If the latitudes touch the cylinder, it is secant case. When the cylinder upon which the sphere is projected at right angles to the poles, the cylinder and resulting projection are transverse. When the cylinder is oblique to the sphere, it is oblique projection. Similarly the projections are classified as tangent, secant and oblique for concial and azimuthal projections (Fig 1.8).

Universal Mercator Projection (UM) is a type of cylindrical projection.

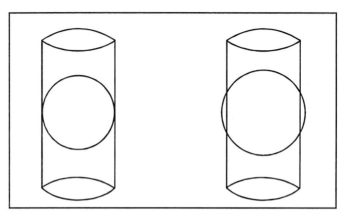

Figure 1.8: Cylindrical Projection

1.7.2 *Universal Transverse Mercator* (UM) *Projections*

UTM was established in 1936 by the International Union of Geodesy and Geophysics and adopted by the US Army in 1947. UTM is being used by many national and international mapping agencies for topographic, thematic mapping and for referencing satellite imagery.

Earth is divided into 60 zones with 6 degrees of longitude and numbered eastward, 1 to 60, beginning at 180° (West longitude)

As the distortion at pole will be greater this is used for the zone between 84° N to 80° south latitude.

Universal Polar Stereographic projection (UPS) is used at poles

Further along latitude, each zone is divided further into strips of 8° latitude

To reduce the distortion across the area covered by each zone, scale along the central meridian has been reduced to 0.9996

The unit for coordinates is in meters

eastings (x) expresses displacements eastward

northings (y) express displacement northward

The central meridian is given an easting of 500,000m. The northing for the equator varies depending on hemisphere. In Northern hemisphere, the equator has a northing of 0 m and for southern hemisphere, the equator has a northing of 10,000,000m

UTM is commonly used for georeferencing and is consistent. Zone number, easting and northing are used for georeferencing. But across zone boundaries, one should be careful while georeferencing to do the work accurately.

1.7.3 Conical projection

The transparent sphere is kept inside the cone and light glows. The features of earth surface are transferred to the cone and the cone is cut vertically and opened. This is conical projection (Fig 1.9).

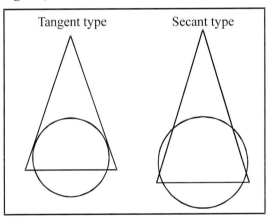

Figure 1.9: Conical Projection

Examples are Alber's conical equal area projection with two standard parallels, Lambert conformal conic projection with two standard parallels and equidistant conic projection with one standard parallel.

1.7.4 Azimuthal projection

The transparent sphere is kept inside a square wall and lights glows. Earth features are transferred on to the walls. Here only half of the features are visible. Other half is shown on other side. Hence projections is possible for only half of the earth (Fig 1.10).

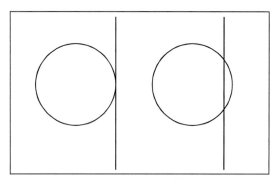

Figure 1.10: Azimuthal Projection

Importance of map projection

The available maps are in different map projections. To use all these maps for a particular applications, it is required to bring all the maps into a common or single map projections.

During map projections, angles, areas, directions, shapes and distances become distorted.

Any map projection will distort one or other properties. No single map projections is available to avoid all the distortions. Hence one distortion will be minimum and other will be more. It is important to select a suitable map projections for any particulate area in the earth.

A map projection is used as a coordinate system. Central meridian and central parallel meeting point is the origin of the coordinate system. Fig 1.11 show the coordinate system used normally. Here all the quadrants posses negative value except NE quadrant. Handling negative value is not convenient for many users and hence user can specify false easting and false northing values. To avoid such negative value larger x and y value will be given so that negative value in other quadrants become positive and working with GIS comfort for the users. Choosing false easting and false northing is optional in GIS softwares.

Many books are available for map projections and users may use those books for further reading.

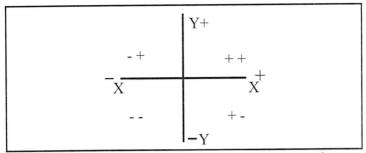

Figure 1.11: Cartesian coordinate system used for false easting and false northing

1.8 REPRESENTATION OF EARTH FEATURES IN GIS

All the objects available in earths surface are represented in GIS by points, lines or polygons.

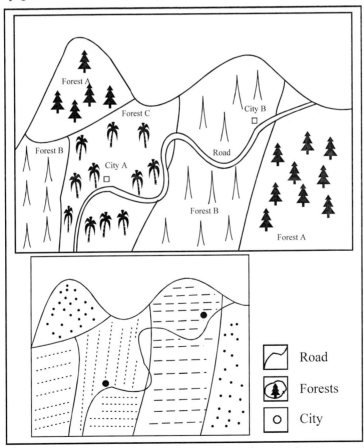

Figure 1.12: Real World view and map view

Fig 1.12 show the real world view & the same is represented as map.

Point: It has a location with x and y in map without spatial dimension as length or width. Water sample location, cities are represented as point data

Line: This is a one dimensional feature with length but no width. Rivers, roads, water distribution system and telecom lines are represented as lines.

Polygon: This is a enclosed feature with area and perimeter. Lakes, land parcels, crop areas are represented as polygons.

Attribute data are the data which gives description about the locations. They are either Continuous data or discrete data

Temperature, rainfall are varying in earth continuously and represented as continuous data. If they are classified at intervals then they are discrete data.

Fig 1.13 show both the continuous data and discrete data. The lines with numbers are discrete in nature. But in between lines pixels are available and they represent continuous features. For example 70 and 75 contours are available and they are discrete data. But between 70 and 75, pixels are available with their numbers ranging between 70 and 75. Hence they are continuous data

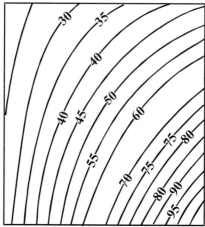

Figure 1.13: Discrete data and continuous data.

Spatial Data

It consists of location, shape and size in earth. A village boundary is represented as polygon in map. The village is having certain shape of irregular polygon with definite area. As all these are mappable, they are called as spatial data. But the village name, revenue of village are description about village are called as non spatial or aspatial data.

Line, point and polygon are spatial data.

Non-spatial information comprises attributes.These are quantitative characterization of the spatial data and descriptive information about geographical features.

1.9 COMPONENTS OF GIS

A GIS database should consists of location of objects(point, line and polygon), attribute and topology (relationship among objects) (Fig1.14).

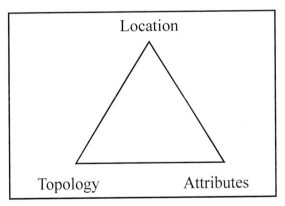

Figure 1.14: Relationship of GIS data

GIS infrastructure

Hardware, software, data, methodology, application, organization and people are together form GIS infrastructure (Fig 1.15)

Keyboard is used to enter attribute data and description about spatial data. Scanner is used to scan paper maps and satellite imageries. Digitizer is used to digitize paper maps. CDs, floppies, zip drives which consists of stored data digitized already available and incorporated into GIS.

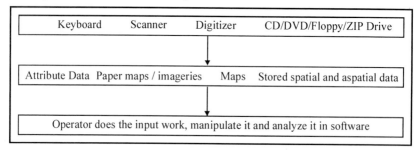

Figure 1.15: Components of GIS

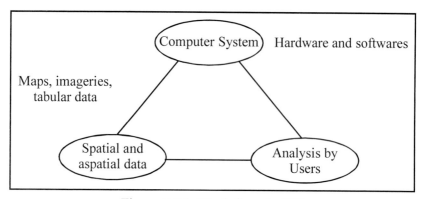

Figure 1.16: Work flow in GIS

Attribute data are collected by survey or from various agencies. Paper maps and satellite imageries are obtained from government and private organizations. Fig 1.16 shows the work flow in GIS.

The data are created by GIS experts and analysed in computer. The analysed results are shown in computer monitor as images, graphs, maps, tables, statistics and tables. Otherwise they are printed in printer or plotter and shown it to administrators for planning and decision making. The results can be stored in floppies, CDs or zip drives to use in other computer. In network environmental data is shared among various GIS experts.

Input Devices

The presence of low cost computers with high configuration made it easier to use GIS even at house level. Floppies of 1.44 MB storage, CDs of 650 MB size, DVDs of 4.5 GB disk and external zip drives of 250 MB are available for storing data.CD and DVD drive are available as read and write and as they posses high memory capacity, are good for GIS projects. Hard disks with 80 GB are available and this will be a good hard disk. 256 MB is available and also 512 MB RAM is available and this will good for GIS processing. The speed to write and read is also being improved highly. Zip drives are connected in USB drive and it is easier to access and write the data in such drives. Personal computers and workstations are becoming synonymous. GIS softwares are working in UNIX, Windows and LINUX systems.

Output Devices

High end scanners, digitizers are available for inputting data. The GIS results are shown as maps, tables, images, graphs in computer monitor or printed using plotters and printers with very good quality.

GIS softwares

Following is the list of currently available GIS softwares for various applications. The analytical capability of the softwares are different. Some softwares are used only for simple GIS analysis while some other could do advanced analysis. Depending on the project undertaken the right software may be purchased for usage. All the GIS softwares posses some basic GIS functionalities. If one is in need of advanced analysis, software vendors provide the software as moduels which will work in the basic software.

ArcGIS 8.2 (From ESRI)consists of :

ArcInfo 8.2, ArcView 8.2, ArcView Spatial Analyst, ArcView Network Analyst, ArcView 3D Analyst, ArcView Geostatistical Analyst.

28   *GIS: Fundamentals, Applications & Implementations*

- Mapinfo from mapinfo corporation
- Microstation Geographics for Bentley
- Geomedia from Intergraph
- AutoCad Mao 2000 from Autodesk Inc.,
- GRASS from US Army Corps of Engineers
- GRAMM++ for CSRE, IIT, Mumbai
- IDRISI from Clark labs
- ILWIS from ITC, Netherland
- Geomatica from PCI Geomatics
- Atlas GIS
- CARIS
- SYCAD

Some of the softwares are raster based and some are vector based and conversion between these formats are possible in some of the softwares. These softwares consists of their own data format and export and import of data from one software to other software is possible with or without loss of data. An voluntary organization called Open GIS Consortium (OGIS) is trying to propose a common GIS format so that data interoperability is possible without difficulties.

1.10 DATA FOR GIS

Photogrammetry

Photogrammetry is the collection of measurements of ground features from aerial photographs. Various techniques are available for interpretation and collection of data from aerial photographs. Large scale maps are produced from aerial photographs. Digital Photogrammetry workstations are nowadays useful in interpretation of data accurately and quickly.

Field Data Collection

Tapes were used for measuring distances. Clinometers or level screws were used to find the gradients and elevation differences. Compass were used for taking approximate readings and theodolites were used for taking accurate angle measurements. The data are processed at lab and corrections applied. But recently due to advent of Electronic Distance Measuring devices(EDM), it is possible to take accurate readings, storing them, transferring them to computer, using packages for interpretation.

GIS is intelligent if strong database is available. Data creation is the big task and 90% of time is spent for data creation only and analysis is done just few mouse clicks.

Paper maps from Survey of India (SOI), NATMO, corporations, cadastral records, satellite imageries from various satellites, radar images, aerial photos, GPS observations and various attribute data of spatial and aspatial in nature are the input for GIS.

Components of a GIS Projects

GIS data comes from many sources. Data from various sources are used to create maps with same size and resolution and taken for result. All the maps are stored as a project (work space is a synonymous term). Hence a project consists of many individual layers. Individual layers may be worked and can be saved. Fig 1.17 show the components of a GIS project.

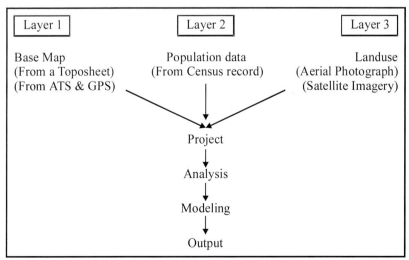

Figure 1.17: Components of a GIS project

Layers in GIS

A GIS project consists of one layer or combinations of many layers. Table 1.2 show some of the layers that are used for various projects. From this, some different layers are derived. Soil layer consists of various types of soils. Permeability type may be given to the soil type and a new layer called soil permeability layer is created. Fig 1.18 show the layer concept in GIS.

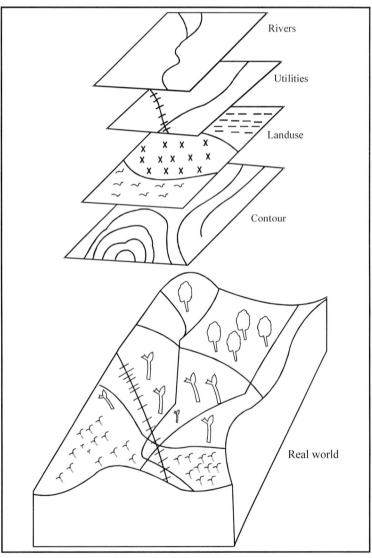

Figure 1.18: Layers in GIS

Geographic Information System 31

Table 1.2: Layers used in GIS for different applications. The layers are used individually or in combination with others.

Layer	Description
Soil	Types of soil
Road	Type of road, intersection, signals *etc.*,
Annotation	Text information of features
Geology	Rock types
Forest	Types of forest
Landuse/landcover	Landuses and natural land cover
Rail road	Railway track with signals, bridges
Sewer line	Sewer systems with connections
Water distribution line	Water pipelines with valves, sumps, reservoirs *etc.*
Contour	Elevation data
Slope	Slope with percentage or angle
Aspect	Slope direction
Hill shade	Shaded map with particular altitude of sun from particular angle
Parcel	Land with ownership details
Flood	Area inundated by flood
Village, city and other administrative boundaries government/private organizations	All types of administrative boundaries of government/quasi
Pollution area	Pollution of land, water, and air
Bore well/open well	Well types with owners
Route map for rail, air, water	Route for buses, planes and ships
Telecom line	Telecom line with details
Electricity line	Electricity connections
Crops	Types of crops
Groundwater potential area	Area for groundwater extraction

Contd....

Table 1.2 Contd...

Layer	Description
Geomorphology	Features like hill, pediment *etc*
Groundwater level	Groundwater level above msl
Places of worship	Temples, mosques, churches
Hotels, shops etc	Hotels, shops, dams *etc.* point locations represented
Drainage	Drainage map with orders
Wetland	Boundaries of wetland

GIS could handle data from many disciplines and hence it became popular to be used in diversified disciplines.

Applications of GIS and Resource Management   **2**

2.1 INTRODUCTION

GIS is a tool that can be used in varied disciplines ranging from civil engineering to business. Remote sensing data is often brought into GIS to have a complete data set in many natural resource application areas. Full capability of GIS is not used widely. Many a times people use GIS only for digitising maps and to do simple query on maps. But it can also be used in modeling, simulation and forecasting, only then one may regard GIS technology is utilised to its full potential. Some GIS softwares handle simple problems for location based services while other complex softwares posses advanced modules to do all types of spatial, network and 3D analysis. Integration of softwares used in natural resources study are being interfaced with GIS to create scenarios.

Remote sensing and GIS technology is useful in the following fields:

— Protection of environment.

— Water resources management.

— Urban planning and transportation planning.

— Watershed management.

— Surveying.

— Coastal zone management.

— Natural disaster management.

34 *GIS: Fundamentals, Applications & Implementations*

- Terrain characterisation and evaluation.
- Infrastructure.
- Utilities.
- Agriculture.
- Forestry.
- Geosciences.
- Demography.
- Biodiversity.
- Wildlife *etc.*

2.2 CIVIL ENGINEERING

Surface water and sub-surface water resource estimation is carried out when the regular hydrology data along with satellite imagery and attribute data are used. When survey using total station or GPS is done, those data shall be brought into GIS and contour, area, slope, aspect, hill shade view, 3D view and perspective view is created. These maps will be useful for siting buildings in a desired location. Also cut and fill volume calculations are carried out. When these maps are loaded in GIS based computer which is integrated with GPS, cutting and filling in ground can be carried out accurately which is now not possible in field in developing countries. Water level variation analysis in regions are possible in GIS. Site selection of dams, reservoirs, roads, new industrial area, railway route, MRTS, site selection for artificial recharge of groundwater are carried out using GIS. Maps like landuse/landcover, contour, soil, rock structures, base maps *etc.*, are required for these analysis. Attribute data like rainfall, water level, weather data *etc.* are required. When satellite imagery for two dates are available for a city then the development between the two dates is identified. When this data is used along with other collateral data, regions for further development can be identified alongwith areas with saline water intrusions can be identified.

2.3 ENVIRONMENT

Spatial information technology application is enormous in the field of environment because most of the environment data

Applications of GIS and Resource Management 35

are spatial in nature. Solid waste disposal, siting of industries, Environment Impact Analysis (EIA), Sedimentation in lakes, dams, river courses, coastal zone management, water, air and land pollution *etc.* are effectively carried out in GIS. Polluted areas are found and remedial measures can be suggested to eliminate the pollution. Also the pollution level can be monitored. Also if an industry is letting out its effluent in a river within 1km radius, then the industry may be shifted. Soil erosion can be estimated. Nowadays EIA study is made using GIS. Coastal zone management is possible using GIS. Any activity within Coastal regulation (CRZ) zone may be shifted based on the buffer created in GIS along the coast line. Presently many species are endangered and many areas in the world are coming under biodiversity hotspots. GIS is a very good tool to find the relation between the terrain characteristics and living conditions for the species. Accordingly decision may be made to conserve such risk zones.

2.4 DISASTER MANAGEMENT

Earth surface is facing one or other disasters frequently. Natural disasters like cyclones, landslides, earthquakes, floods, avalanches, droughts, forest fires mapping are being done at regular time interval. Disasters like earthquakes cannot be forewarned but its likely occurrence is predictable. In such zones a seismic resistant buildings may be built. Movement of cyclones are predicted and warning is being given. Areas susceptible for landslides, avalanches are identified and developmental activities may be reduced in such areas. Based on rainfall and water resources availability areas susceptible to droughts are identified and relief measures are done. Movement of forest fire is tracked using GIS and such areas may be warned for the risk of forest fire and thus timely action can be taken to safeguard the resources.

2.5 EMERGENCY PLANNING

During artificial disasters like bombing, leakage of poisonous gases, bursting of chemical plants, facilities required for relief measures, routing of vehicles and resource mobilization

36   GIS: Fundamentals, Applications & Implementations

is carried out. But spatial data along with other data is not available in most of the cities of the world except a few cities of developed countries. In near future GIS data will be available for most of the big cities of the world so as to face challenges and resource management.

2.6 GEOLOGY

Mineral, water, oil and rock resources are essential for the development of a country. Besides the site selection of dams, tunnels, road are important. Earth posses many weak zones that are faults which are responsible for earthquake. GIS is largely used in mapping rock structures like lineament, fault, dip, strike, fold, joint, are mapped. Groundwater availability is determined based on the soil, rock permeability and joints in the rock along with various geomorphic units. Crustal deformation study using GPS and with ground data are used to find earthquake vulnerable zones. Areas with oil resource is identified. When the bursting of volcanoes, are that will be affected may be identified and preventive measures may be taken. All such studies are done in GIS environment. 3D view of the mining and underground is created to understand the sub-surface features.

2.7 UTILITY MANAGEMENT

Utilities like water distribution system, sewer system, telecom lines, power lines, gas pipe lines, transportation network can be mapped in GIS and can be maintained well. Fig 2.1 shows the commonly available utilities. All these utilities are digitized and the attributes of the components are added. All these utilities are georeferenced to earth. Hence if an utility engineer is interested to repair an object in a utility on road side, it is possible to use a GPS to identify the object location without digging here and there. The exact point is identified with GPS and rectification work is done. Also adding a new line may be planned at shortest path form the existing facility. Accidents are plotted in a road map and the reason for an accident may be known and proper action is taken to avoid such accidents in future.

2.8 FORESTRY

Forest resource is important to have a balance in the ecosystem. Trees with their inventory, age of the trees, health of the trees are recorded and monitored. Also density of forest, timber estimation, deforestation and afforestation are monitored. Wildlife corridors for the movement of animals are constructed which are based on GIS. Also tracking of wild animals using GPS integrated with GIS made the scientists understand the wild animals behavior in a much betterway.

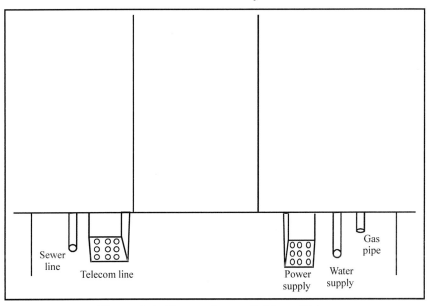

Figure 2.1: Commnly available utilities

2.9 AGRICULTURE

Crop water requirement, crop disease monitoring, crop yield estimation, stress of crop are done using remote sensing and GIS. Precision farming is the new area where agricultural activities are modernized. The agriculture field does not possess nutrient content uniformly. But fertilizers are used uniformly in the entire field. But in reality some areas already have more nutrient contents while other areas do not have enough nutrient contents. As same amount of fertilizer is used throughout the field, this results in wastage of fertilizers, in huge amount. In developed countries, maps with soil nutrient content is available in a computer with GIS with GPS integration which is available in tractor. When the tractor is moving in the field, the sensor available in the tractor will release less amount of fertilizer where more nutrient content is available and more amount of fertilizer where less amount of nutrient content is available in the soil. Soil erosion is a major problem in hilly regions and also in gently slope areas. Erosion leads to nutrients erosion from soil. Hence such areas shall be identified and preventive measures may be taken. Shifting cultivation is planned systematically so that it will lead to sustainable development. Crops and their health is identified from satellite imageries and water is released to cultivated areas according to the health of the crops.

2.10 BUSINESS

GIS application is enormous in setting up a facility in suitable locations. Developed countries possess all digital data like road, census, landuse and all other related facilities. Hence integrating the available data is possible in GIS and according to the result site for a new bank, a new shop or any other facility is broughtand designed. Real time monitoring of goods, call taxis, and trucks is possible. When the GIS based map is available for cities, tourists shall impose spatial queries and an answer is found. Adding a facility is planned according to social status of the population.

2.11 LAND INFORMATION SYSTEM AND CADASTRAL RECORDS

This is a sub-set of GIS. This consists of surface and sub-surface features for a region with regard to natural resources and are parcel based. Facilities of the parcels are also added. Mineral, soil, agriculture and all other resources of land are given in LIS. Cadastral records are created and maintained for taxation, land registration and to have history of the land dealings. Presently many departments are having control over selling a land property. But if all the records are available and are created then, any one government office may take all the responsibility of the land transaction and duplication or cheating will thus be reduced. If such cadastral are available in a web access mode, then public will have transparency about the land records and they shall do their transactions. Also as the database will be available centrally it is possible to access throughout the state. Cadastral records of such type will come into existence in near future even though already cadastral records are computerized to some extent.

2.12 MILITARY

Terrain characterisation is important for military applications. Contour and 3D view, landuse/landcover, infrastructure facilities are important for defense. Viewpoints between two or more locations may be useful for strategic planning to execute an effective attack. Timely supply of logistics and other required support can be provided efficiently.

GIS Techniques and Nature of Data

3

3.1 INTRODUCTION

Geographic Information System handles large volumes of data. It is important to know the nature of data, instruments from which the data is collected and represented in computers. GIS data is either spatial or aspatial. Data is an important component in GIS. Digital data are available for some projects. but for many projects, input data may not be available. Now-a-days, data is supplied by many government, private organizations and also available on internet. If data for a particular project is not available, then the user has to collect the data from satellite imageries, GPS and other instruments and maps from various sources.

Spatial Data

All locations in the earth surface are spatial data. Location of a city, location of a river, location of a lake are represented in earth by their latitude or longitude in global coordinate system or as x,y coordinate in Cartesian coordinate system. Spatial data consists of location, shape and size in earth. A village boundary is represented as a polygon in map. The village is having certain shape of irregular polygon with definite area. As all these are mappable, they are called as spatial data. But name of the village, revenue of a village and description about a village and called as non-spatial or aspatial data.

42 *GIS: Fundamentals, Applications & Implementations*

Line, point and polygon are spatial data. Fig 3.1 shows the different forms of spatial data representing the real world data in map based on the nature of data and its distribution.

In the (Table 3.1), the addresses are located in a map based on their address location in terms of x,y or latitude or longitude. Name of the owner and land value are aspatial component.

Table 3.1: Spatial and Aspatial data

Address	Name of the owner	Land value in Rs/sq.ft.
13, VK Road, Coimbatore	Kannan	300
101, Avanashi Road	George	500
02, Ram Nagar	Abbas	400

Spatial Data Relation

Distance: Between two points distance is measured

Distribution : Occurrence of feature in a map. Distribution of certain trees in a forest may indicate some relation with other natural conditions.

Density: Density of population of cities or districts or some

Aspatial data: Descriptions of spatial data. Name of a city, name of river, name of a lake are aspatial data.

Following are the data types used in day to day life and also common in GIS.

Boolean : 0 or 1

Nominal : Any names

Ordinal: 0 to α

Integer: Whole number - α to + α

Real: Real numbers form -α to + α

Topological whole numbers
Position are two types namely

1. Absolute - xyz, latitude and longitude
2. Relative - direction, proximity.

GIS Techniques and Nature of Data   43

Figure 3.1: Various forms of Map representing the real world data.

Data types should be entered when creating attribute tables. The higher level data types are categorical and numerical. Categorical data types includes nominal and ordinal scales while numerical includes ratio and interval scales. Measurements scales are important for data display. Numerical data is important in data analysis. Character strings are used for representing nominal and ordinal data. Integers and real numbers are used to represent interval and ratio data.

Examples of data types

Nominal	Rock type, soil type, water body, urban area, paddy field - No Mathematical operation
Ordinal	First, second ranking - median and percentile - large, medium and small cities- No Mathematical operation
Interval	15° to 30 ° C- correlation analysis, regression analysis, Arithmetic mean
Ratio	Based on a meaningful or absolute zero value -All mathematical operations for real numbers allowed
Time	Time duration - monthly rainfall, flow characteristics Time frequency - river gauge reading 3 times a day

3.2 DATA STRUCUTRES

Real world data are represented as either raster or vector data in computer. Both structures have their own advantages and disadvantages. During the initial development of GIS it was difficult to process both vector and raster data in the same GIS software. Softwares were developed to handle either raster or vector data. But now-a-days, it is possible to do analysis on raster and vector data in many of the existing softwares. Also conversion from vector to raster (rasterisation) and raster to vector (vectorisation) is be easy by present day softwares.

3.2.1 Raster Data Structures

All the scanned maps and documents are in raster form . All digital satellite imageries are in raster format. In raster form the objects are represented as pixels (picture element). 3D representation of pixels is called as voxels.

GIS Techniques and Nature of Data 45

Raster data does not maintain true shape. Five water sample locations may be represented in a single map. Then the map is converted into grids (pixels of certain row and column). Cells with water sampling locations will have value of 1 and remaining cells will have a value of 0 (Fig 3.2). When data is not available for an area a class called no data is available. No data means that data is not present and this is not 0 value.

1	0	0	0	1
0	0	0	0	0
0	0	1	0	0
0	0	0	0	0
1	0	0	0	1

Figure 3.2: Representation of water sample location in a raster map

But in vector to represent the water sample locations, they are located as points of x and y coordinates. No 0 value of no data is present. These map look like a normal map one could see in daily life (Fig 3.3).

Raster data consist of resolution. This is minimum mapping unit. To represent a land parcel of 10m x10m size, then the resolution of raster cell should have the resolution of 10m x 10m.

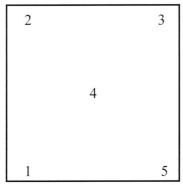

Figure 3.3: Vector representation location

If the raster cell is with 15m × 15m resolution, then the land parcel will not be represented.

If two points occur closely then they may be represented in a single cell. Fig 3.4 show both raster and vector representation of the points. Here 2 points located at closer distance among other two points and the result is the representation of only 9 points. It is not possible to distinguish point data like bore well locations and street lamp post for a city in a single raster map. Two maps are needed to present, one map for bore well locations and other for street lamp locations. Raster cells may be coded if even a cell is partly covered by a feature. Other method is if a cell centre is filled with some feature the the feature will represent the object.

During vector to raster conversion of landuse map, a single cell may consists of 10% coverage of agriculture, 60% by building and 30% by open land. Then if we ask for to fill up the same cell

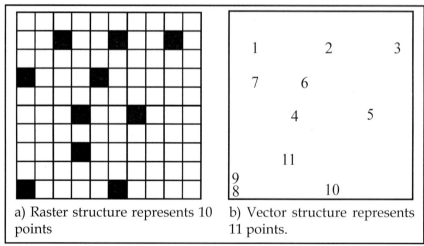

a) Raster structure represents 10 points

b) Vector structure represents 11 points.

Figure 3.4: Raster and Vector representation.

where a dominant feature is there, then the entire cell is filled with building. Such things may happen for some more cells. If one is interested to find the area occupied by buildings for a region, the result will be wrong. Options are also available to fill up a cell completely, even if only 10% of a cell is filled by a particular landuse. Raster cells are good in representing square and rectangular shapes but not good in representing circular and triangle shapes.

Various types of raster data structures are available. Fig 3.5 show the simple raster data structure. Raster data compression methods like run length coding ,block coding,chain coding and quadtree structure.

										10, 10, 3
1	1	1	1	1	1	1	1	2	2	1, 1, 1, 1, 1, 1, 1, 1, 2, 2
1	1	1	1	1	1	1	2	2	2	1, 1, 1, 1, 1, 1, 1, 2, 2, 2
1	1	1	1	1	1	2	2	2	2	1, 1, 1. 1, 1, 1, 2, 2, 2, 2
1	1	1	1	2	2	2	2	2	2	1, 1, 1. 1, 1, 2, 2, 2, 2, 2
1	1	1	1	3	3	3	2	2	2	1, 1, 1, 1, 3, 3, 3, 2, 2, 2
1	1	1	3	3	3	3	2	2	2	1, 1, 1, 3, 3, 3, 3, 2, 2, 2
1	1	3	3	3	3	3	3	2	2	1, 1, 3, 3, 3, 3, 3, 3, 2, 2
1	3	3	3	3	3	3	3	3	2	1, 3, 3, 3, 3, 3, 3, 3, 3, 2
3	3	3	3	3	3	3	3	3	3	3, 3, 3, 3, 3, 3, 3, 3, 3, 3
3	3	3	3	3	3	3	3	3	3	3, 3, 3, 3, 3, 3, 3, 3, 3, 3

Figure 3.5: Simple raster data structure

Simple raster data structure

Here all the cell consists of numbers. All the cell values are represented as such and this occupy more memory space. To avoid more memory space requirement other types raster compression methods developed.

Run Length coding

Fig 3.6 show the run length coding. Here number of rows and number of columns and number of features (or information) available is represented in first row. Second indicates 0 value for all cells (10 cells) of the first row in image. but second row of image consists of 0 value for 1 cell, 1 value for 6 cells and again 0 value for 3 cells. Likewise all the 10 rows are represented. As all the cells with their information are represented, the requirement of memory is large.

0	0	0	0	0	0	0	0	0	0	10, 10, 1
0	0	0	0	1	1	1	1	0	0	0, 10
0	0	1	1	1	1	1	1	1	0	0, 4, 1, 4, 0, 2
0	0	1	1	1	1	1	1	1	0	0, 2, 1, 7, 0, 1
0	0	0	1	1	1	1	1	1	1	0, 2, 1, 7, 0, 1
0	0	0	0	1	1	1	1	1	0	0, 3, 1, 7
0	0	1	1	1	1	1	1	1	1	0, 4, 1, 5, 0, 1
1	1	1	1	1	1	1	1	1	0	0, 2, 1, 5, 0, 3
1	1	1	1	1	1	1	1	1	1	0, 1, 9, 0, 1
1	1	1	1	1	1	1	1	1	1	1, 10
										1, 10

Figure 3.6: Run Length coding

Block coding

Fig 3.7 show the block coding. Here the information available in blocks are represented. Three different types of information are represented in 3 blocks with column and row position with number of cells.

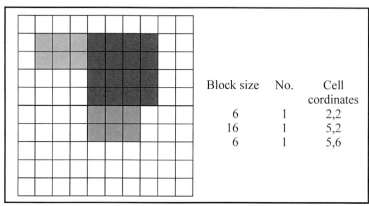

Figure 3.7: Block Coding

Chain coding

Fig 3.8 represent chain coding. Cells with information are starting and closing. The chains are stored. Here the first cell for the chain starts at column 2 and row 7. From these cells, 6 cells available in north, 4 cells in east, 3 cells in south, 2 cells in east, 2 cells in south, 6 cells in west. Now the chain is closed.

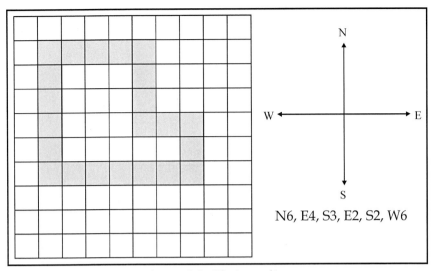

Figure 3.8: Chain coding

Quadtree

Fig 3.9 show the quadtree structure. Here information is available partially in block number 3 and no information is available in blocks 1,2 and 4. As information is available in block 3 it is further sub divide into 4 blocks. Here block 31 is completely filled with information. Block 32 consists of information and hence it is further sub divided into 4 blocks. Here only 321 and 322 blocks consists of information. The same image is represented as tree in the figure below the quadtree image. Quadtree is the recent development in raster data structures with very good applications. Here the raster blocks are divided into smaller and smaller cells. If a raster map is available with 10m resolution, then all the cells are representing the features with 10m x 10m size. It is not possible to represent a 1m object in the 10m resolution. But in quadtree structure it is possible to represent 1m resolution object in 10m resolution map.

Advantages of Raster Data Model

1. Simple data structure
2. Good for overlay and overlay with satellite imagery is easy

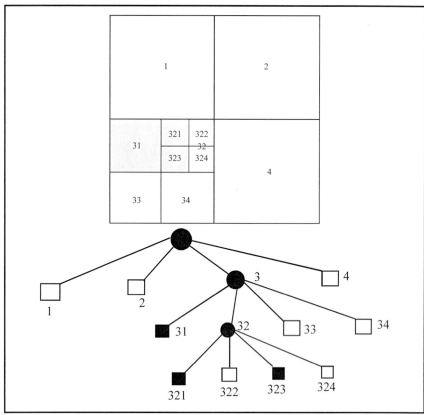

Figure 3.9: Quadtree data structure

3. Satellite imagery is used without any difficulty or conversion as satellite imagery are available in raster format
4. Various types of spatial analysis possible
5. Simulation is easy because each pixel is in same size and shape
6. Same set of pixels are used for several variables
7. Simpler when own programming is done
8. Raster handles transition well
9. Analysis, simulation and modeling, easy

Disadvantages of Raster Data Model

1. More memory space required
2. Errors in estimating perimeter and shape

3. Network linkages are difficult to establish
4. Topology is difficult
5. Less accurate and low resolution
6. Resolution affect the data quality, usage of larger cells will omit the recognition of important small structures

Raster data analysis are overlay, DEM, slope, aspect. 3D view, cut and fill volume calculation, buffering, cost path analysis, simulation, modeling and change detection.

3.2.2 VECTOR DATA STRUCTURES

The data is represented in Cartesian coordinate system with x and y coordinates. Two types of vector data structures are available. They are spaghetti file and topology structures files.

Spaghetti file

Here the x and y coordinates are stored for points, lines, and polygons based on the digitization direction and no relationship among them is available. Fig 3.10 show the spaghetti structure.

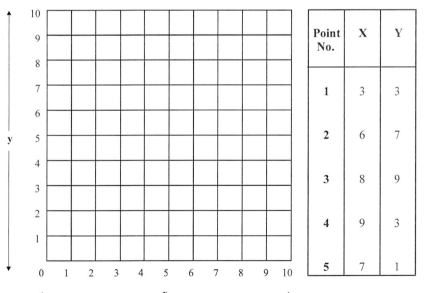

Figure 3.10: Spaghetti size

Topology vector data structure

When a road network analysis is done, the software is not able to find the route. Hence the user has to give from node, to node, overpass, underpass, road cross, left turn, right turn *etc.*, Normal digitization produce spaghetti model without topology. It is important to build topology to have the relationship of objects. The relationship is among points, point to line point to polygon, line to line, line to point, line to polygon, polygon to polygon, polygon to point and polygon to line.

Topology is useful to find shortest path from one location to other. Occurrence of a lake within 10 km radius of a dense forest for serving water to wildlife. Fig 3.11 show the topological vector data structure.

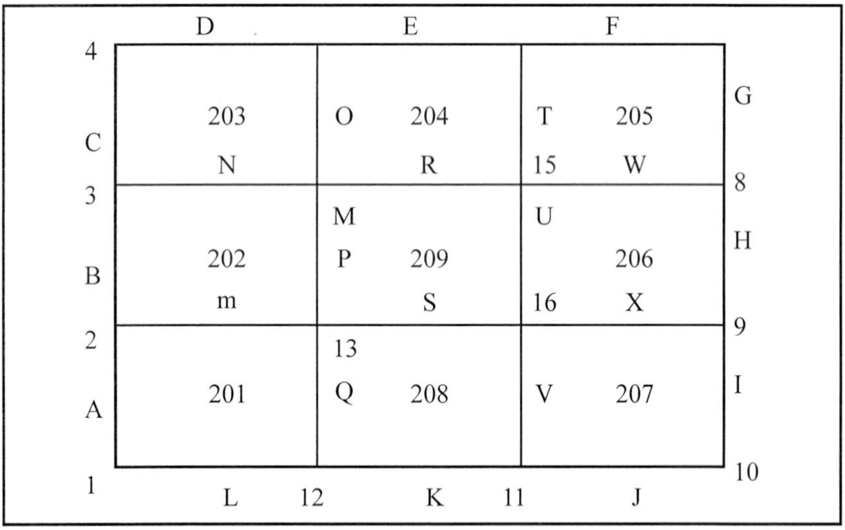

Figure 3.11: Topological structuring of complex areas

This consists of nine polygons numbering from 201 to 209. Here polygon 209 is called as island polygon as it is surrounded on all sides by polygons. Numbers 1 to 16 are called as nodes. A to X are called as arcs or segments or chains. A topologically structured area consists of node file (Table 3.2), Arc file (Table 3.3) and polygon file (Table 3.4). Attribute data may be added to the polygons. Here for a polygon's left polygon and a right polygon exists. Hence they are topologically structured. Table 3.5 shows the attributes for the polygons.

Advantages of Vector Model

1. Less storage, high resolution, represents real world feature as such with pleasing to eye
2. Compact Data Structure
3. Accurate graphics
4. Network linkage is good (topology)
5. Used for administrative boundaries

Table 3.2: Node File   **Table 3.3:** Arc File

ID	X	X
1	0	0
2	0	18
3	0	22
4	0	30
.	.	.
.	.	.
.	.	.
.	.	.
16	22	18

ID	Start node	End node	Left polygon	Right polygon	area
A	1	2	Outside	201	250
B	2	3	Outside	202	200
C	3	4	Outside	203	250
D	4	5	Outside	203	250
.	.	.	.	.	.
.	.	.	.	.	.
.	.	.	.	.	.
..X					

Table 3.4: Polygon file   **Table 3.5:** Polygon attribute file

ID	Chain list
201	AMQL
202	BNPM
203	CDON
204	ETRO
.	
.	
.	
.	
.209	PRUS

ID	Crop	Soil Nutrient
201	Paddy	High
202	Sorgum	Low
.	.	.
.	.	.
209	Paddy	High

Disadvantages of Vector Model

1. Complex data structure
2. Simulation is difficult because each unit has different topological form
3. Overlay of many maps difficult
4. Spatial variation could not be well represented
5. Simulation and modeling is difficult.

Vector data analysis: Buffer, vector overlay, distance measurement, network analysis.

Hardwares for Data Input and Output

Scanners, printers, plotters and digitizers are designated based on the handling capacity of paper size. The different size of hardwares commonly used are :

A0 size 34"	x	46"
A1 size 22"	x	34"
A2 size 17"	x	22"
A3 size 11"	x	17"
A4 size 8.5"	x	11"

If a scanner / printer / plotter / digitize can handle an A0 size sheet then the hardware is called as A0 size scanner or printer / plotter / digitizer respectively. All the hardwares are supplied with a driver operating CD to install the hardware devices.

There are many commercial brands of scanner, printer, plotter, digitizer available

3.3 RASTER DATA INPUT

Scanners

Based on the mode of feed their are two types of available scanners. They are flat bed scanners and drum scanners. In flat bed scanner, the map is kept inside and it is scanned. If the paper is large which is used in big size scanners, then the image is scanned row by row and all of them are tailored into a single image. In drum scanner the map is inserted and the map is moving along with the drum during scanning. Map area which is in contact with drum is scanned and fed into computer. A map can be scanned in available scanning softwares and saved in any one of the programs like Adobe Photoshop, Adobe, Microsoft Word, Microsoft Powerpoint, Paint *etc*. The Image Reddy image may be thus scanned and saved in any of the raster formats.

Scanning converts paper map into raster digital data. It consists of 0 and 1.0 is background and 1 is the map feature. For scanning image a resolution of 300 dpi is required (dots per inch). The scanner may be a monochrome or color scanner and accordingly the image is scanned in black and white or colour.

Black and White maps are scanned as 1 bit monochrome. To obtain grayscale image, 8 bit is adopted while 24 bit is used for colour image.

To get 24 bit true colour, scanning is done in 3 colour (RGB). Each colour has 8 bit integer number with range of 0-255.Roughly 24 bit colour image can have 16.7 million colours.

Pixel depths are given as follows.

4 bit	16 colours
8 bit	256 colours
16 bit	32,768 colours
24 bit	16.7 million colours.

On screen digitization or heads up digitization is expensive when compared with automatic raster to vector conversion.

Scanner will not enhance the quality of the output because the output is relied directly on the source or the base map. If source map contain errors that error will be carried out in the digital data also. The quality of digitized data depends on the quality of the source maps and effort taken by the operator to digitize the map accurately. Paper maps are subjected to shrink and expand due to the adverse effect of temperature and humidity. Due care should be taken while scanning the maps which were already subjected to some level of distortion.

If spatial resolution increases by two times, total size of the 2D raster image will increase by 4 times, because the number of pixels gets doubled in x and y directions.

Complete raster to vector conversion is possible. It includes image acquisition, preprocessing, line tracing, text extraction (OCR), shape recognition, topology creation and attribute assignment.

Automatic digitization

The map to be digitized is set for proper resolution. Then preprocessing is done to remove annotations, stains, wrinkles in the map. Map is scanned at 3 pixel width. Later thinning is carried out for the central pixel line only. This is vectorized and all the errors are removed. Labels are added, topology is built, georeferencing is done and the map is brought into GIS for analysis.

Problem in automatic digitization are the break lines in contours for labels, annotations. Many roads are available and softwares do not know how to trace the line. The operator has to control the softwares at every stage, while, stains, wrinkles and converging contours pose problems. Stains, wrinkles and converging contours pose problem.

Digitization may be carried out using Cad softwares like AutoCad or Microstation. Also scanned images are inserted into GIS and digitized. Few automatic raster converson sfotwares like R2V is available where the raster map is automatically vectorised.

Automatic raster to vector conversion

Few automatic raster to vector conversion softwares are available.

They automatically or semi automatically digitize the maps. They supports various image formats like TIFF, GeoTIFF, BMP etc., at 1 bit bi level (0 and 1), 8 bit gray scale and color images (4 bit, 8 bit, and 24 bit). Also supports satellite imageries. Georeferencing of images are possible. It can also import/export Arc view shape file, ArcInfo Generate, Mapinfo MIF/MID and AutoCad DXF.

In one command the entire scanned image is vectored within a second. Otherwise to trace a contour, the cursor is kept on contour, the line is selected automatically and digitized. Also heads up digitization is possible. All types of vector editing is possible. Many layers can be created. Bilinear and Delaunary triangulation method is used for geometric transformation. The digitized vector layer can be georeferenced. Maps can be edge matched. 3D view is also created. Man is able to pick up 1/40 inch in a map. But scanners can scan and produce an image of

200 to 800 dpi (dots per inch). Automatic vector conversion softwares can draw a centre line or boundary line on these 800 DPI resolutions and they will produce accurate digitization than manual digitization. In manual digitization, the operator's hand may be shaking, sloppy, distracted, tired, lazy, impatient, *etc.*, and they will produce inaccurate data.

The transformations are carried out by bilinear and triangulation method. Bilinear method corrects global distortion while triangulation corrects local error. Bilinear method is useful when less number of control points are available and the original map has got very less local distortion.

Triangulation method require at least 8 control points, 4 at 4 corners and 4 inside the image and all these eight are evenly distributed. Triangles are generated throughout the image and all the local errors are rectified. Also the control point locations are not moved from its original location. The map is ready and can be brought into GIS for analysis.

3.3.1 Remote Sensing

Remote sensing means acquiring information about objects without having in physical contact with the object. Human eye is an example for remote sensing device. Like this satellites revolving at higher orbit take images of earth at various spectral and spatial resolution.

Aerial photography was the first remote sensing data, providing visual imagery of landscape on films. In recent decades, sophisticated electronics and hightech sensors have been developed that gather data from various parts of electromagnetic spectrum that are invisible to human eye.

Remote sensing satellites take images in the visible and infrared regions of EMR (Electro magnetic spectrum). Everything in nature has its own unique distribution of reflected, emitted and absorbed radiations. Fig 3.12 shows the energy interaction in earth's atmosphere. These energies are different for different objects. At temperature above absolute zero every objects radiate electromagnetic energy. Electromagnetic spectrum is a dynamic form of energy that propagates as wave motion at a velocity of 3×10^{10} cm/sec.

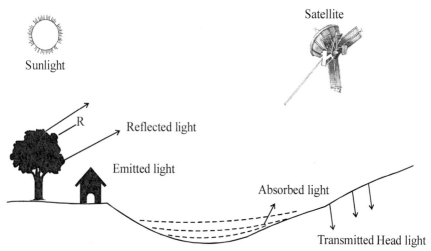

Figure 3.12: Energy Interaction in earth

On the basis of source light used remote sensing can be categorised into two. Active remote sensing means passing known source of light to the ground and recording the reflected radiation. Radar satellites employ this method. But most of the satellites working is based on passive remote sensing principle. Here solar energy is the source. When the solar energy reaches on the earth a part of it is reflected back in the atmosphere itself, while some part is reflected by the ground objects and some part is refracted. Sensors in the satellite records the energy at various spectral levels. Charge coupled devices (CCDs) are used to sense the energy.

Radiation form the sun, when incident upon the earth's surface are transmitted into the surface or absorbed and emitted by the surface. The EMR, on interaction, experiences a number of changes in magnitude, direction, wavelength, polarization and phase. These changes are detected by the remote sensor and enables the interpreter to obtain useful information on the objects of interest. The remotely sensed data contain both spatial information (size, shape and orientation) and spectral information (tone, color and spectral signature).

The interaction of EMR with atmosphere is important. The information carried by EMR reflected/emitted by the earth's surface is modified while traversing through the atmosphere The atmospheric constituents scatter and absorb the radiations. Rayleigh scattering, Mie scattering and non-selective scattering are caused by air molecules, smoke and clouds respectively. The phenomenon of refraction in the atmosphere also affects the EMR. Ozone, CO_2 and water vapour absorb the EMR passing through atmosphere in certain spectral bands.

Electromagnetic Energy Spectrum

Electromagnetic spectrum is expressed as wavelength of energy ranging from very short (cosmic ray on the left) to very long (television and radio waves on the right). Visible light is the very narrow portion of the electromagnetic spectrum meaning that human eye see's only a small part of that spectrum whereas remote sensing equipment can obtain information from much greater portion of the spectrum. Each portion on the spectrum has particular advantage in remote sensing. Fig 3.13 shows the electromagnetic spectrum.

Middle range of electromagnetic spectrum contain wavelengths that are visible to the human eye (visual) ranging from blue to red. Ordinary camera film is sensitive to this spectrum producing standard aerial photography. Electronics and some films captures, next larger wavelength just outside the visual, called the near infrared. The near infrared reveals information that are not available in the visible spectrum, such as plant health and stress. Visual infrared can show damaged vegetation before it is apparent to the observer on the ground. Thermal infrared sensors detect very small temperature differences and display themes on imagery. Warmer water is brighter than cooler land. This is useful in thermal pollution monitoring. For example industrial effluent can be analysed in terms of heat characteristics. Urban heat sources can be detected easily and landscape thermal mapping can reveal geologic and vegetation differences. Thermal data is important in fire fighting.

60 GIS: Fundamentals, Applications & Implementations

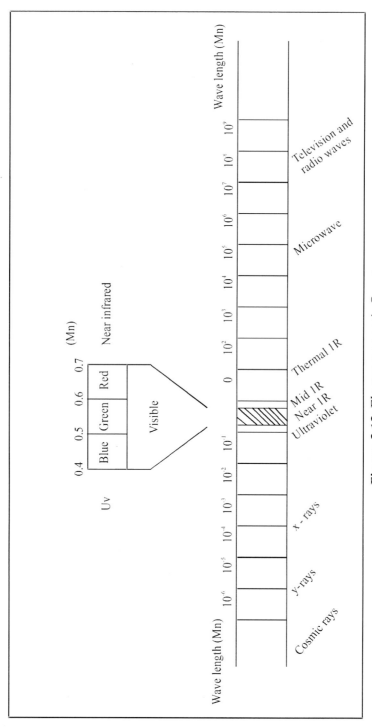

Figure 3.13: Electromagnetic Spectrum

Radar microwave data are long wave, producing land and water information much different from that of the visual region of the spectrum. Radars useful are penetrating in cloud cover, and can help to map topography in humid cloud covered tropical region where aerial photography has been unsatisfactory. It can also detect some types of vegetation and can detect subtle geologic features such as faults. Microwave is useful in some oceanographic application, such as oil spill mapping and monitoring sea ice distribution.

Some of the remote sensing satellite available for resource management are given in Table 3.5

Table 3.5: Remote Sensing satellites with their characteristics

Mision	Country	Year of Launch	Imaging Mode	Resolution (m)
Landsat 1,2,3,4,5	USA	1972, 75 78,82,84	MSS TM	79 30
SPOT1,2,3	France	1986,90, 1993	PAN HRV	10 20
IRS 1A&1B	India	1988, 91	LISSI&II	72 (LISS I) 36 (LISS II)
IPS P2	India	1994	LISS II	37x32
IRC 1C/ IRS 1D	India	1995/97	LISS III WiFS PAN	23.5 (VNIR) 69 (MIR) 188 (WiFS) 5.8 (PAN)
NOAA -9 10,11,12,14,15	USA	1985,86 88,91,94,97	AVHRR	1100 km
ADEOS	NASDA/ Japan	1997	MSS PAN	8 16
TERRA-ASTER	USA/Japan	1998	MSS	15
IKONOS	Space Imaging, USA	1999	PAN MSS	0.82 3.2
KOMPSAT	KARI, South Korea	2000	PAN	6.6
EROS A1	Israel	2000	PAN	1.8
Quickbird2	Earthwatch, USA	2001	PAN MSS	0.61 2.44
Orb View 3	OrbImage, USA	2002	PAN MSS	1/2 4

Resolution of Images

Resolution is defined as the ability of the system to render the information at the smallest discretely separable quantity in terms of distance (spatial), spectral, temporal and radiometric.

Spatial Resolution: Scanners spatial resolution is the minimum size of the ground segment sensed at any instant.

Spectral resolution: It is a measure of both the discreteness of the band widths of the sensor and the sensitivity of the sensor to distinguish between grey levels.

Radiometric resolution: Dividing the total range (B to W) of the signal output into a large number of just discriminable levels so as to be able to distinguish ground features differing only slightly in radiance or reflectance.

Temporal resolution: Repetivity of the satellites.

Visual Image Interpretation

An interpreter studies remotely sensed data and attempts through logical process to detect, identify, measure and evaluate the significance of environmental and cultural objects, patterns and spatial relationships. It is an information extraction process. For this purpose hard copy of black and white and colour images are used.

Size and Shape

Numerous components of the environment can be identified by their shapes. Size of the objects like length, breadth and area are important in image interpretation. A playground may be in oval shape. Regular arrangement of square shape may be indication of agricultural field. The approximate size of objects can be judged by comparisons with familiar features (*e.g.* road) in the same scene.

Tone

Different objects emit or reflect different wavelengths and intensity of radiant energy. Such differences are recorded and are useful for discriminating the objects. A dense forest is represented in darker tone while a paddy field is represented in lighter tone.

Pattern

Repetitive pattern of both natural and cultural features are quite common, which is fortunate because much image interpretation is aimed at the mapping and analysis of relatively complex features rather than the more basic units of which they may be composed. Such features are farms and orchards, alluvial valleys and coastal plains.

Texture

Texture is an important image characteristic closely associated with tone in the sense that it is a quality that permits two areas of the same overall tone to be differentiated on the basis of micro tonal patterns. Common textures are smooth, rippled, mottled, lineated and irregular. Two rock units may have the same tonne but different textures. Deciduous forest may have coarser texture while coniferous forest may have fine texture.

Site

At an advanced stage in image interpretation, the location of an object with respect to terrain features of other objects may be helpful in refining the identification and classification of certain picture contents. For example, the combination of one or two tall chimneys, a large central building, conveyors, cooling towers and solid fuel piles point to the correct identification of a thermal power station.

Resolution

Imageries posses different types of resolution and they should be considered while interpretation.

Stereoscopic Appearance

When the same feature is photographed from two different positions with overlap between successive images, an apparently solid model of the feature can be seen under a stereoscope. Such a model is termed as a stereo model and the three dimensional view provided can be thus used in aiding interpretation. This valuable information cannot be obtained from a single imagery.

Stereo images are used for large scale mapping and used for the selection of suitable sites for infrastructure projects.

GROUND TRUTH

Ground truth refers to any verification of mapped data against true ground conditions.

Digital Image Processing

Commonly available Satellite Image Processing Softwares are

- ERDAS IMAGINE
- ENVI
- EASI/PACE
- ER MAPPER
- IDRISI
- ILWIS
- GRAMM++
- IDRISI, ILWIS and GRAMM++ consists of both image processing and GIS component.

Remotely sensed data is available in CD ROM, magnetic tape and in floppy diskette.

A digital remotely sensed image is typically composed of picture elements (pixels) located at the intersection of each row i and column j in each k bands of imagery. Digital imageries are in raster format. Associated with each pixels a number known as Digital Number (DN) or Brightness Value (BV) that depicts the average radiance of a relatively small area within a scene. A smaller number indicates low average radiance from the area and high number is an indicator of high radiant properties of the area. When represented as numbers, brightness can be added, subtracted, multiplied, divided and in general subject to statistical manipulations that are not possible if an image if presented only as photograph. Now-a-days due to the availability of advanced and cheapest hardwares and softwares, digital image processing is widely used rather than visual interpretation by photograph.

IMAGE RECTIFICATION

This involves the initial processing of raw image data to correct for geometric distortion, to calibrate the data radiometrically and to eliminate noise present in the data. This is a pre-processing operation.

Image Enhancement

It involves techniques for increasing the visual distinction between features in a scene. New image is created from the original image to obtain more information form the image. The operations are level slicing, contrast stretching, spatial filtering edge enhancement, spectral ratioing, principal component analysis and intensity-hue-saturation colour space transformations.

Image Classification

This operation is to replace visual analysis of the data with quantitative techniques for automating the identification of features in a scene. This involves the analysis of multispectral image data and the application of statistically based decision rules for determining the land cover ideality of each pixel in an image. Both the unsupervised and supervised classification methods are adopted.

Supervised Classification

Supervised Classification is a technique for the computer-assisted interpretation of remotely sensed imagery. The operator trains the computer to look for surface features with similar reflectance characteristics to a set of examples of known interpretation within the image. These areas are known as training sites.

Unsupervised Classification

Unsupervised Classification is a technique for the computer-assisted interpretation of remotely sensed imagery. The computer routine does this by identifying typical patterns in the reflectance data. These patterns are then identified by undertaking site visits to a few selected examples to determine their interpretation. Because of the mathematical techniques used in this process, the patterns are usually referred to as clusters.

Aerial Photography

Aerial photographs are used to take aerial photos for large scale mapping. Fig 3.14 show the geometry of aerial photograph.

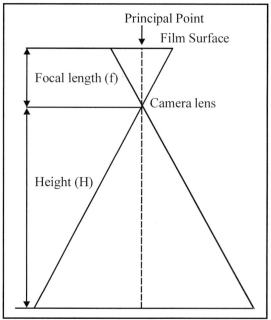

Figure 3.14: Geometry of Aerial Photography

3.4 VECTOR DATA INPUT

This is the process of converting paper (analog) map into digital data. Digitization is done using Digitizer (Fig 3.15). Digitizers are available in various sizes ranging from A0 to A4. Digitizing tablet consists of inbuilt electronic wire mesh. A puck is available which is similar to mouse which will pick up a point a line or a polygon. The digitizer working area consists of x and y axis as in a graph. If a point is clicked inside the working area, its coordinates are shown or digitized. Large digitizers possess 0.003 cm accuracy.

GIS softwares posses modules to interact with digitizer. Topology could be built during digitization itself. A set of control points are used to register the map first and then digitization is carried out. For digitizing line, point or stream mode is used. In point mode user is clicking on various points and they are stored.

But in stream mode, if the user tracing one a line, points are automatically added at predetermined time/distance specified by the user. Point mode is good for digitizing features with more or less straight lines while stream mode is useful when the line is often curved. Also discrete or continuous mode is available for digitizing a line or polygon. In discrete mode, the user specifies the intersection of nodes. But in continuous mode, the user does not see's the intersection and digitizes as it wishes but later the GIS software will identify such nodes. Digitizing line and polygon are more or less same but the difference is a label which is given for each polygon.

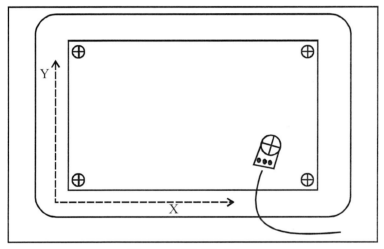

Figure 3.15: Digitizer

Digitizing line twice around a polygon may give duplicate line. This duplicate line shall be made into one line by setting proper tolerance. Slivers (polygon resulted due to overlay) are also created and this is also removed by setting proper tolerance. After digitization the map is printed and checked for errors. The printed map is divided into many small blocks. Each block of digitized map is compared with the original map. If an error is noticed, they may be noted and later rectified. This procedure is much useful to identify and rectify the errors without omission. Otherwise if the entire digitized map is compared with original map, it is not possible to observe the errors due to the operator's attention on the entire map.

Geometric Transformation

A digitized map is read as the source map. Hence after digitization the maps should be referred to real world coordinates. This is done using a set of control points. Geometric transformation is converting the raster map or digitized data into real world coordinate or a projection system using a set of control points and transformation equations.

Different types of geometric transformations are available.

Equal area transformation: This allows rotation of rectangle and preserve its shape and size.

Similarity transformation: This method allows rotation of rectangle and preserve shape and does not preserve size.

Affine transformation: This allows angular transfor-mation but preserves the parallel lines.

Projective transformation

This allows both the angular and length distortions but the rectangle is converted into irregular quadrilateral.

Topological transformation

This preserves topological properties of objects but convert a rectangle into circle. Most of the packages use affine transformation widely even though it is possible to do similarity and projective transformations. Which is used to correct aerial photographs.

Affine Transformation

Affine transformation preserves parallelism but allows differential scaling, rotation, scaling, skew and translation (Fig 3.16).

Rotation rotates the x and y axes from its origin. The image is shifted to a new location by translation. This is skewed into a parallelogram by skewing and its x or y axis is expanded or reduced by differential scaling.

Geometric correction uses affine transformation or polynomial transformation. Affine transformation involves rotation, translation and scalng. As satellite images are oriented north east, affine transformation is good. GCPs are used to run affine transformation. This will run with as et of control points and root mean square error(RMS) is examined. If the RMS error is more then different control points are given till the expected RMS is obtained. Polynomial equations are used to geoerefence an image also called as rubber sheeting or warping. Here the image is subject to differential scaling and rotation. Second order polynomial is used. Map projection or geometric transformation involves data resampling to fill the values of the new grid created from the existing grid. Nearest neighbor, bilinear interpolation and cubic convolution are the three methods used for resampling. In nearest neighbour, nearest cell value is assigned to the new grid from the original grid. In bilinear interpolation, four nearest cell values from old grid is weighted and given to the new grid cell. Cubic convolution method uses nearest 16 cell values from original cell value and weighted and given to the new grid. Such resampling is also used for creating DEM from contours.

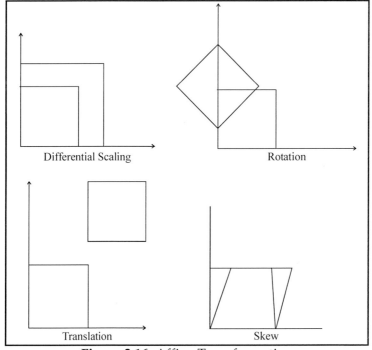

Figure 3.16: Affine Transformation

70   *GIS: Fundamentals, Applications & Implementations*

Commonly available digitizers are
1. Altek series
2. Calcomp series
3. Hitachi series
4. Summagraphics series

Following are some of the digitization terms:

Vertex: a point or a line with known coordinates

Dangle: an error, line which crosses instead of meeting

Node: the end point of line

Puck: mouse like devise attached with digitizer on a digitizing tablet . It consists of cross hair which will be carried out through a line or point or polygon to be digitized

Tablet: The digitizing table which consists of many wires embedded in it for detecting the magnetic field generated by puck.

Snapping: A process by which two lines will join automatically if snap is set.

3.4.1 Global Positioning System (GPS)

Global positioning system (GPS) are used for the calculation of latitude, longitude and altitude for many users. GPS has 3 segments; the space segment, the user segment, and the control segment. The space segment consists of 24 satellites, each in its own orbit 20,200 kilometers above the earth's surface. The user segment consists of receivers, which can be a hand held type or mounted in vehicles. The control segment consists of 5 ground stations, located around the world which monitors the proper functioning of GPS. The Master Control facility is located at Schriever Air Force Base (formerly Falcon AFB) in Colorado. The monitor stations measure signals from the satellites which are incorporated into orbital models for each one. The models compute precise orbital data (ephemeris) and satellite clock corrections. The Master Control station uploads ephemeris and clock data to the satellites. The satellites then send subsets of the orbital ephemeris data to GPS receivers over radio signals

GPS receivers can be held, carried or installed on cars, trucks, aircraft, ships. These receivers detect, decode, and process GPS satellite signals. The typical hand-held receiver is just like a cellular telephone. The information about positioning and navigation can be read on the display or transferred to other equipments (*i.e.* computers).

The basis of GPS is "triangulation" from satellites. To "triangulate," a GPS receiver measures distance using the travel time of radio signals. The whole idea behind GPS is to use satellites in space as reference points for locations on earth. By very accurately measuring distance from three satellites one can triangulate his position anywhere on earth. Fig 3.17 shows that four satellites are required to have latitude, longitude and altitude.

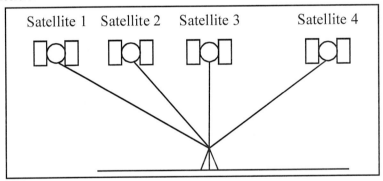

Figure 3.17: GPS Survey to locate location and altitude of a place.

Three types of GPS are available. First is Standard Position Service (SPS) and is available to everyone. Horizontal position accuracy is within 100m and vertical position is within 300m. Second one is Precise Position Service and at earlier times it was available only to US military and other authorized users but now it is available for everyone. The accuracy in Horizontal position is within 22m and vertical position is 27.7m. The third is Differential GPS (DGPS). It is based on the principle that most of the errors seen by GPS receivers in a local area will be common errors. A base station GPS receiver (master) is established over a known location and a second receiver (rover) is located at the position to be determined. Both GPS receivers track the same satellites. Any discrepancy between the position indicated by the master receiver and its actual position can be attributed to system errors, atmospheric errors or selective availability (the chief sources of positional error). The necessary corrections can be calculated and applied to the rover's calculated position as well. Some surveying quality differential GPS systems are reported to achieve sub 0.01m accuracies. Mapping and Surveying applications typically require considerably more accuracy than is available from either Standard Position Service or Precise Position Service although they are both more than adequate for most navigation purposes. Differential GPS allows considerably higher accuracy

GPS service is offered by US Department of defense called NAVSTAR GPS and the second is developed by Russia called as GLONASS. NAVSTAR (Navigation Satellite and Time Ranging) GPS is used widely. The GPS constellation consists of 24 satellites in 6 orbital planes. This provides accurate time information through rubidium and cesium clocks. GPS system was actually developed for military purpose but by finding its enormous use this has been made available for civilian purposes. European Union has also planned its GPS mission called the Galelio.

GPS Signals

The positional information from the satellites are transmitted to the surface through two L band frequency called L1 and L2 with the frequencies of 1575.42 MHz and 1227.6 MHz. respectively. The most important aspect of the position information that is obtained from the GPS satellite is its reference system. The WGS 84 Coordinate System is a Conventional Terrestrial Reference System (CTRS)

Datum is the most important function for GPS surveys. Everest datum is used for India and its adjacent countries for measurement. Horizontal and vertical datum is available and often horizontal datum is used. A horizontal geodetic datum consists of latitude, longitude, an azimuth of a line (direction) to some other triangulation and flattening of the selected ellipsoid for computation. Four satellites are needed for measuring latitude, longitude and altitude for a location.

GIS Techniques and Nature of Data

Applications of GPS

Real time kinematic GPS has opened floodgates to new applications. GPS is also being increasingly used for machine guidance. One can actually have a digital map of the job site and display it in the bulldozer with control down to several centimeters. It enables machine operator to go directly to the right coordinates. GPS are attached with machines. GPS can also contribute significantly to accurate mapping of wetlands. GPS can also serve as a watcher on structures or natural features where ground movements are common. GPS receivers placed on a bridge, dam or hill top linked by a modem in a home base can measure the movements quickly through GPS position changes and can send an e-mail or pager message or warning if danger threatens. The positional data from GPS can be given as input into GIS softwares and contouring, cross sections, cut and fill volume calculation and alignment of roads, pipelines, tunnels *etc.*, can be carried out easily.

Geodetic measurement provides very accurate determinations of positions of points on the earth's surface and GPS is highly useful for this measurement. Tectonic plate movement study requires millimeter accuracy but navigational accuracy is a meter to few centimeters. Scientists, sportsmen, farmers, soldiers, pilots, surveyors, hikers, delivery drivers, sailors, despatchers, fire-fighters, and people from many other walks of life are using GPS in many ways while makes their work more productive, safer, and sometimes even more easier.

GPS is very much useful in surveying. High end GPS are necessary for survey applications. Time requirement for GPS surveying is very minimum compared with other conventional survey equipments. Also It is possible to survey an area for industry or for other purpose aerially. A flight fitted with GPS can survey an area with in a shorter time span. Vehicle monitoring during construction work, alignment of canals, tunnels or road can be carried out accurately with GPS.

GPS Data

Differential correction means a known base station is surveyed and established accurately. One more station within 500 km of base station can accurately find its position relative to base station. This is differential GPS. Selective Availability Degradation to broadcast orbit and dithering of satellite clock Antispoofing Denial of P code to civilian users.

Types of Positioning

1. Static Single Point Positioning
2. Static Relative Positioning
3. Kinematic Single Point Positioning
4. Kinematic Relative Positioning

Errors in GPS

Following are the types of GPS errors. Software packages will correct these errors at post processing level

1. Satellite Clock (10m)
2. Orbital (100m S/A Active, 5-25m S/A inactive)
3. Ionospheric (50m at zenith)
4. Tropospheric(2m at zenith)
5. Receiver clock (10-100m)
6. Multipath
   C/A Code-50cm to 20cm
   Carrier- upto few cm
7. Receiver noise
   C/A code - 10cm to 2-3m
   Carrier - 0.5 to 5mm

GPS Project Phase

This involves four stages as given below

1. Planning and Preparation
   (i) Validation & (ii) Reconnaissance
2. Field Operation

3. Data Processing
4. Final Reporting

Before going to the field all proper planning should be undertaken. As the project to be undertaken is known, all the required layers and instrumental arrangements may be carried out well in advance. A reconnaissance survey may be carried to have an exposure on the field. Then actual survey may be carried out and data is stored. Now-a-days most of the high end GPS consists of Personal Digital Assistant (PDA) with softwares like ArcPAD to enter the data. Proper map projection and datum may be set for surveying. The data may be downloaded into computer and all editing may be done. Now the layers are ready to import into GIS. Interfacing of GPS with GIS is not at all a problem. Then the data is overlaid with GIS and the result is derived.

GPS Applications
1. Surveying.
2. Alignment of Roads, Canals *etc.*
3. Monitoring sway of buildings, bridges, slopes.
4. Verticality of buildings.
5. Machine control, Geodetic control points, Photo control.
6. Urban Mapping/Planning.
7. Energy-Oil,Gas Pipe Lines.
8. Remote Sensing.
9. Utilities-Sewerage and Telecom.
10. Watershed Management.
11. Railway/Rural Road Mapping.
12. Rural Cadastral.
13. Military Seismic studies (Plate Movements).
14. Forestry/Resource Management.
15. All mapping applications.

76 GIS: Fundamentals, Applications & Implementations

Total Station

Surveying with total station provides millimeter accuracy for terrestrial survey. Surveying has been made easy as it encompasses all survey components like bearings, slope distance, x, y, z distance. Total station consists of instruments, batteries and prisms. When infrared rays are emitted and they reflected back hit the prism kept at a location. Distance upto 5km can be covered using this survey. Laser based instruments are also available which do not require prisms and any object can be focused to obtain the result. The surveyed data taken can be imported into GIS and contouring, cut and fill volume calculations are found.

Database Creation and Analysis in GIS **4**

4.1 INTRODUCTION

Database management system is a computer program to store, search, manipulate retrieve and give the result. Vast amount of spatial data are available and it is necessary to arrange them, store them for retrieval and analysis. Attribute data adds intelligent to GIS. Without database, a GIS will be just a Cad software. Cad softwares do not have database. GIS softwares have their own databases besides it has the potential to link with external database like Oracle, MS Access, MS Excel, Informix, INGRESS, Sybase and dbase. The data available in paper has to be digitally stored in a proper order. Previously the data were stored in a a flat file format where the data was stored without any structure. But to search and retrieve and analyse the database, database structures were developed. There are of 4 types of database structures namely Hierarchical, Network, Relational database and Object orientation.

Tables are made of rows and columns. Rows represent a map feature. Columns represents characteristic of map feature. Row is called a record or a tupule column is a field or an item. Table 4.1 is a table where rows 1, 2 and 3 represent the locations with the soil types. A column soil type represents various soil types.

Table 4.1: Components of a table

S.No	Location	Soil type
1	Gandhipuram	Black cotton soil
2	Sulur	Red soil
3	Mettupalayam	Alluvial soil

78 *GIS: Fundamentals, Applications & Implementations*

ArcInfo uses INFO, ArcView uses dbase, AutoCAD map uses VISION. In georelational model, attribute data and spatial data are stored separately. Both are linked by ID and analysis can be made. But in object orient model, both spatial and attribute data are stored in same the database which saves processing time. Object orientation model is being continuously upgraded.

When digtising a point, line or polygon, a database is automatically created. Required fields shall be added and information about the objects may be given. If the attribute data is stored in any external database, they shall be linked.

If a table is too big, the table can be splited into many smaller tables and they may be added with the working tables in GIS. Two separate tables may linked be a common field, it is called as key.

4.2 DATABASE CONCEPTS

The available four types of database concepts are hierarchical, network, relational and object orientation. Hierarchical database organize data at different levels and contain one to one relationship. In network database, connection exists between tables. Problem with both the hierarchical and network database is complicated and inflexible database.

But most of the GIS use relational database concept. In relational database, connections are related with each other by keys. Relational databases are simple and flexible. Databases can be prepared, edited and maintained separately. When needed, two tables are linked by common ID. This enhance efficient data management and data processing.

4.2.1 Hierarchical database

Fig 4.1 show a hierarchical relationship of tables. A is a parent and A1 and B1 are child. Here a parent has got child. But there is no relation between A and B also A1 and B1. Quadtree is a type of hierarchical type. Here in each department may consists of many sections. Hence this is a tree like structure. Searching is fast and updating is easy in hierarchical database but the disadvantage is the linkage is only vertically but not horizontally and diagonally.

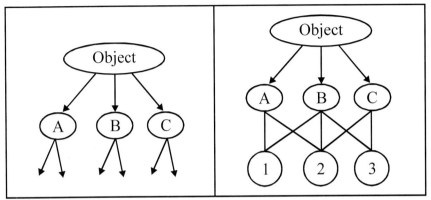

Figure 4.1: Hierarchical database structure

Figure 4.2: Network database structure

4.2.2 Network database

Here the linkage between a parent with children of other parent possible and flexibility exists to some extent. Fig 4.2 show the network database structure. Parent A and B is in relation with child 1.

4.2.3 Object Oriented Model

An Object Oriented model uses functions to model spatial and non-spatial relationships of geographic objects and the attributes. An object is an encapsulated unit which is characterized by attributes, a set of orientations and rules.

An object oriented model has the characteristics of generic properties, abstraction and adhoc queries. Fig 4.3 show the object oriented database structure.

Inheritance relationship is shown by generic properties. It is possible to ask spatial query explicitly without much normal relational queries. Objects, classes and super classes are abstracted well. A type of spatial language is asked in orientation model. Object orientation model exists but not at its fullest capability for GSIS analysis. If one is interested to ask questions like where is Coimbatore? In object orientation, the model is

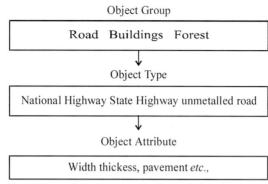

Figure 4.3: Object Oriented database

represented by the way that Coimbatore is present in Tamilnadu. Here latitude and longitude of Coimbatore is not considered but the spatial relationship of Coimbatore represented in Tamil Nadu is explicitly represented.

Data should be stored in a structured manner. If data is not arranged in a proper order then analysis is not easy. An unstructured table is called as a flat file. Table 4.2 is an example for flat file. Fig 4.4 show the Hierarchical database concept. Here two zones *viz* agriculture and built-up land is present. Below this zone, parcel numbers are represented. Below the parcel number, owners are represented. Zones numbers are called as parents and parcel numbers are called a child. Then parcel number is a parent and owner is a parent. Hence parent and their child relationship is established. Hence hierarchy is maintained.

Table 4.2: Flat File

Parcel No.	Owner	Landuse zones
101	Karthick	Agriculture(1)
101	Shyam	Agriculture(1)
102	Shivanna	Builtup land (2)
102	Ragavendra	Builtup land (2)
103	Vivek	Builtup land (2)
104	Peter	Agriculture(1)
104	Mubarak	Agriculture(1)

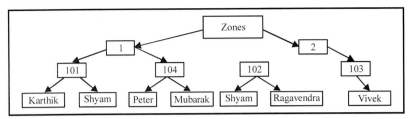

Figure 4.4: Hierarchical Model

Fig 4.5 show the network database concept. Here the parent child relationship is crossed.

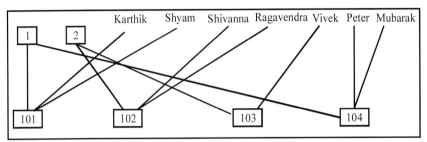

Figure 4.5: Network data Structure

4.2.4 Relational Databse Concept

Types of Relationship In Database

Three types of relationship exists. They are
1. one to one
2. one to many and
3. many to one

Source table is the first table from which data can be exported to other table called destination table. In one to one relationship, one record in destination table is related with one record in source table. In one to many relationship, one record in destination table is related with many record in the source table. In many to one relationship, many records in destination table is related with one record in the source table. (Fig 4.7)

Data are entered by typing or exporting data from other database packages.

Data entered should be verified with original data.

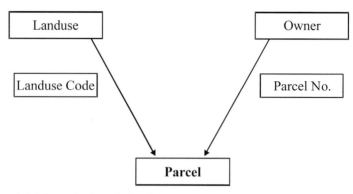

Figure 4.6: The relational database concept.

In relational database concept, a big table is split into many smaller table. Whenever needed the data is called from other table. Table 4.3 to 4.7 show the normalization process of tables.

Table 4.3: Flat file for a zone

Parcel No.	Owner	Owner address	Area in Sq.ft	Zone Code	Landuse Type
101	Karthick	20, VK Road	1500	1	Agriculture
101	Shyam	31, Bharathiar Road	1500	1	Agriculture
102	Shivanna	01, Rajaji Salai	2200	2	Builtup land
102	Ragavendra	22, Peters Road	2200	2	Builtup land
103	Vivek	59, Kamarajar Salai	1200	2	Builtup land
104	Peter	101, Renge Gowda street	10000	1	Agriculture
104	Mubarak	50, Bezant Nagar	10000	1	Agriculture

Table 4.4: Parcel table

Parcel No.	Area in Sq.ft.	Zone Code
101	1500	1
101	1500	1
102	2200	2
102	2200	2
103	1200	2
104	10000	1
104	10000	1

Table 4.5: Address table

Owner name	Owner Address
Karthick	20, VK Road
Shyam	31, Bharathiar Road
Shivanna	01, Rajaji Salai
Ragavendra	22, Peters Road
Vivek	59, Kamarajar Salai
Peter	101, Renge Gowda street
Mubarak	50, Bezant Nagar

Table 4.6: Owner table

Parcel No.	Owner
101	Karthick
101	Shyam
102	Shivanna
102	Ragavendra
103	Vivek
104	Peter
104	Mubarak

Table 4.7: Zone table

Zone code	Landuse type
1	Agriculture
2	Builtup

If a data is required to be imported from other table based on a common field, data can be imported.

4.2.5 Database Design

The data should be easily retrieved without any type of tedious search. Duplication of data (redundancy) may be avoided. Hence redundancy will reduce the cost of the project and increase the processing speed. The created data shall be shared by many people for diversified applications. Also the searching of database by many people at same time should be made possible. Also updating of the database may be made at any time and at any stage. The database should be standardized.

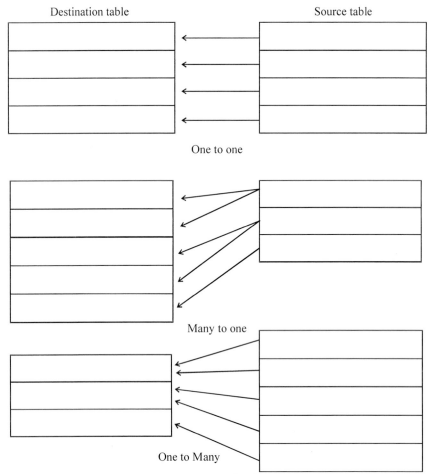

Figure 4.7: Types of relationships

The data available in the database should have followed naming convention, proper decimal points, proper extension, proper symbol.

If one filed called population and other field area is available for a district, then the database should have flexibility to add a new column by dividing the population field by area.

4.3 DATA ANALYSIS

In GIS when a point or polygon or line layer is created a database is also created. Fig 4.8 show the landuse map for an

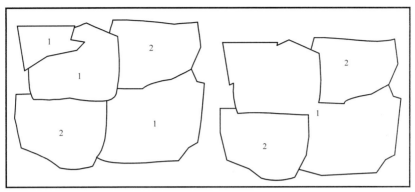

Figure 4.8: Landuse map and recoded landuse map.

area with two landuse types 1 and 2. Recoding is possible in GIS. When recode option is used, features with same types are merged and reflected in the polygon map.

Fig 4.9 show the soil map. Fig 4.10 show the landuse map. Landuse map consists of 5 polygons with three polygons consists of ID No.1 and two polygons with ID No.2. After clipping the clip theme shows the polygon Ids of landuse. Fig 4.11 show the clip theme.

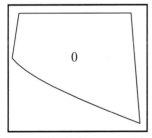

Figure 4.9: Soil map with a polygon ID 0

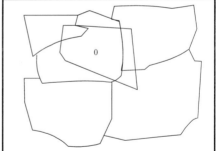

Figure 4.10: Soil map is overlaid with landuse map

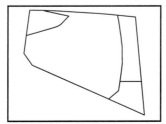

Figure 4.11: Soil map clipped with landuse map

GIS and Time

Many objects change in due course of time. Population, forest cover, soil fertility, urban development, wildlife count, rainfall, temperature subject to change over the years. It is important to see and quantify the changes. Such analysis are called as change detection. Fig 4.12 show the migration of river over the years. Visually the change in river course is identified. Also Fig 4.13 show the deforestation during 1990 and 2005 respectively. Fig 4.14 show the deforestation between 1990 to 2005.

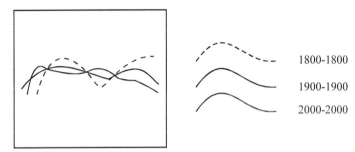

Figure 4.12: Migration of rivers

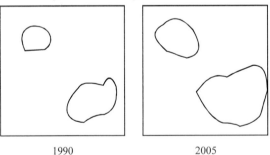

Figure 4.13: Deforestation 1990 & 2005

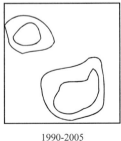

Figure 4.14: Change between 1990-2005

Proximity analysis

Natural objects and also man made objects have neighbourhood relationship. A state highway is connected with national highway. A lower order stream is connected with higher order stream. This is called as connectivity. In traffic density and accident analysis link of state highway with national highway is considered. In watershed study, connection of lower order stream with higher stream is important because water flow from lower order to higher order stream(Fig 4.15). Contiguity is the relationship of an object with other object. Fig 4.16 show the Contiguity of an object with an other object as none, weak, good.

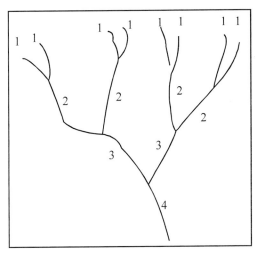

Figure 4.15: Connectivity of Streams

Spatial Relationship of objects

If the real world data are mapped, relationship of features among them will be known and they will be much useful for planning and decision making. Distance, Distribution, Density, Pattern and proximity are some of the spatial data relations. Fig 4.17a show the shortest between the locations A and B. Hence to start from a point and reach the other point within short time such relation is useful. Fig 4.17 b show the distribution of various species. If data like contour, landuse, rainfall, soil, slope , then the distribution pattern for a particular species may be found. Fig 4.17c show the density of plant species. Density is more in some places and less in some places. Such relationship will be due to the combination of various other factors. Fig 4.17d show the distribution pattern of species. Such pattern may be due to particular slope. Fig 4.17e show the proximity of well grown trees near a pond. The healthy plants is due to its proximity to a pond. Fig 4.17f show the old type of building and new type of building. Here the time factor is considered *i.e.* change detection for a city over certain period.

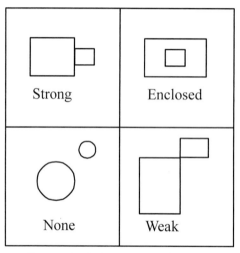

Figure 4.16: Contiguity function

Common data formats

GIS software store both vector and raster data in different formats. For examples when a Microsoft Word is used the document is stored in doc format. An Auto CAD drawing is stored in dwg format. These are proprietary format of the developers. Many GIS softwares are available and they store the data in different formats. Also major organisations have developed softwares and stored in a particular format. Data for a GIS project comes from various sources and it is important to know the data formats. Following are some of the data formats related to GIS.

Database Creation and Analysis in GIS 89

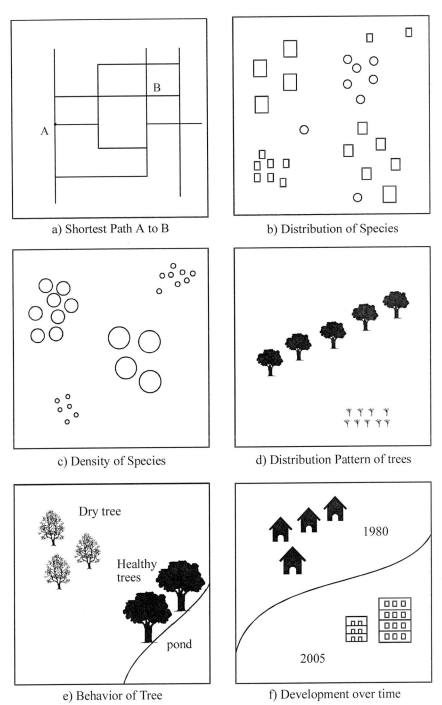

Figure 4.17: Spatial Relationships of Objects/Data

ADS	-	Automated Digitizing system, US Bureau of Land Management
	-	sub system of MOSS
ADRG	-	Arc Digitized Raster Graphics
AMS	-	Automated Mapping System
	-	US Dept. of Interior - sub system of MOSS
BIL, BIP, BSQ	-	Bit Interleaved by Line, Bit Interleaved by pixel, Band Sequential ASCII data description file
BMP	-	Windows bitmap Format
CDR	-	Corel draw Format
CGM	-	Computer graphics metafile Format
Dbase	-	Database file
DIME	-	Dual Independent Map Encoding
	-	US Bureau of Census; replaced in 1990 by TIGER
DLG	-	Digital Line Graph
DTED	-	Digital Terrain Elevation Data
	-	US Defense Mapping Agency Format
DXF	-	AutoCAD Drawing Interchange Format
EOO	-	ESRI's ArcInfo Export Format
EPS	-	Encapsulated Postscript Format
ERDAS	-	ERDAS Proprietary Format
ETAK	-	Mapbase File- digital Street Map (for USA only)
JPEG	-	Joint Photographic Expert Group
Gif	-	Graphics interchange Format
GIRAS	-	Landuse/Landcover data - US Geological Survey format for landuse, landcover
GRASS	-	Geographic Resource Analysis Support System
IGDS	-	Interactive Graphic Design Software
	-	Intergraph's Microstation Format (Design file)
IMAGINE	-	ERDAS Proprietary Format
ISIF	-	Intergraph's Standard Interchange Format
MIDAS	-	Map Information Assembly Display
	-	US dept. of Agriculture Soil Conservation
MOSS	-	MOSS export File- ASCII files readable by US Dept. of Interior - public Domain GIS MOSS

NTF	- National Transfer File-Ordance Survey Format
PCX	- PC Paintbrush Format
Pct	- Macintosh PICT Format
PNG	- Portable Network Graphics Format
RLC	- Run Length Encoded - for monochrome scanned images
SDTS	- Spatial Data Transfer Standard-US Federal Information Processing Standard 173
Shape file	- ArcView Format
SIF	- Standard Interchange Format - used before editing and printing
SLF	- Standard Linear Format US defense Mapping Agency Format
Sun Raster	- Sun Raster file
TGA	- Targa Format
TIFF	- Tagged Image File Format - desk top publishing Format
TIGER	- Topologically Integrated Geographic Encoding and Referencing File- US Bureau of Census
VPF	- Vector Product Format - by US Defense Mapping Agency
WMF	- Windows Metafile Format
WPG	- Word Perfect Graphics Format

Metadata

This provides information about spatial data. Metadata gives information that area of coverage, data quality, and data currency. Also how to use the data. Metadata should have the following information.

1. Identification information – about data, title, area covered and currency.

2. Accuracy like positional and attribute, consistency, methods used to produce data and sources of the data.

3. Raster or vector data

4. Map projections, datums coordinate system and resolution

Data interoperability

If vector data is available in one format some GIS softwares will take the data as such but few softwares will loose some information because if incompatibility.

For example Arcview can import AutoCAD DXF file, Microstation DGN(Design) file, Arc Info's interchange file and Mapinfo's MID file. Geomedia Professional can import data from Microstation, AutoCAD, ArcView and ArcInfo.

But now Open GIS consortium has been formed and it has laid rules to have open format.

TIGER, VPF and DLG file can be read by many GIS packages.

Paper sizes used in plotters

Paper size	Dimension
ANSI A	8.5" x 11.0" (216 mm x 279 mm)
ANSI B	11.0" x 17.0" (279 mm x 432 mm)
ANSI C	17.0" x 22.0" (432 mm x 559 mm)
ANSI D	22.0" x 34" (559 mm x 864 mm)
ANSI E	34.0" x 44.0" (864 mm x 1118 mm)
ANSI F	28.0" x 40.0" (711 mm x 1016 mm)
ISO A4	8.3" x 11.7" (210 mm x 297 mm)
ISO A3	11.7" x 16.5" (297 mm x 420 mm)
ISO A2	16.5" x 23.4" (420 mm x 594 mm)
ISO A1	23.4" x 33.1" (594 mm x 841 mm)
ISO A0	33.1" x 46.8" (841 mm x 1189 mm)
ISO B4	9.8" x 13.9" (250 mm x 353 mm)
ISO B3	13.9" x 19.7" (353 mm x 500 mm)
ISO B2	19.7" x 27.8" (500 mm x 707 mm)
ISO B1	27.8" x 39.4" (707 mm x 1000 mm)

Data for GIS from various mapping agencies, satellite images providers, attribute data supplied by many organisations besides the data captured by GIS users.

Satellite imageries

The data are in the form of Maps, data from GPS and other Surveying Instruments, which collect data like rainfall, temp., reports tables and human input.

Following Steps are involved in GIS workflow for a Project.

(i) Project identification

(ii) Data input

(iii) Editing

(iv) Georeferencing

(v) Projection

(vi) Data conversion

(vii) Topology

(viii) Analysis

(ix) Result

(x) Decision making

When GIS data like point, line and polygon is created it consists of an ID for the features and with a column to add data. Table 4.10, 4.11 and 4.12 show the table created during a GIS project work.

Table 4.10: Data Point layer

Well No.	Location
1	Gandhipuram
2	RS Puram
3	Kovai Pudur
4	Peelamedu

Table 4.11: Line layer

River No.	River name
1	Noyyal
2	Cauvery
3	Bhavani
4	Amaravathi

Table 4.12: Polygon layer

Landuse No.	Name of landuse
1	Builtup Land
2	Forest
3	Waste Land
4	Agriculture area

Following brands of printers/plotters are available in the market.
1. Calcomp Series
2. IBM Laser Printers
3. Epson
4. HP
5. TVS
6. Wipro
7. Toshiba

Labeling the Map

If the name of the objects are stored in database, it is possible to label the object automatically or manually by clicking on the object. All GIS consists of zoom in, zoom out zoom to full extent, zoom to active theme, zoom to selected features. Panning is also possible. In GIS many layers are created and all of them are stored as single project file. Otherwise the layer may be stored as a single layer file. The layer added may be put in visible or invisible by just clicking on the check box. Also information for an object is found just by clicking a tool available called identify tool. The map produced may be printed in portrait or landscape. Distance between two location may be measured. Area of a polygon may be found. The colour of the layers may changed by a single mouse click. A particular city in a big city map is identified by find button. Also cities within a particular distance may be found.

GIS consists of operators like, <, > and, or not *etc.*, for querying various layers.

Data Compression

GIS data are larger in volume especially with raster data. It is important to reduce data volume for effective storage and easy analysis. Data compression programs like Winzip,PKzip and Gunzip are used for compressing voluminous data. Text documents are compressed for 10 - 15 times while graphic images are compressed to only less than 5% of its original volume. An image scanned with simple raster data structure require more storage space while the same file scanned in run length code will reduce the data volume to higher extent. Even during scanning, the scanning resolution of 150 dpi require less space while compared with scanning with 300 dpi or more. But when the numbers in mage are frequently changing, compression efficiency will be less.

Errors in GIS and Data Output

5

5.1 INTRODUCTION

A GIS database without single error is reliable for decision making. There are various possible sources of errors in GIS. If an error is not removed during scanning, the final output will also have that error. Besides whatever the work has been carried on wrongly scanned map it will add to errors because the source map itself's erroneous. This is like garbage in garbage out proverb. If wrong data is taken for analysis it will only result in wrong output. Hence it is important to remove error at every stage of GIS work. It is important to locate the sources of errors so that due care may be taken while handling a project.

5.2 ORIGINAL SOURCES OF ERRORS

All the equipments which give attributed data and spatial data will give errors. Before going to field or before collecting data all the related instruments should be calibrated properly. Also the instrument should not cross the permissible error. Besides these while using data from diversified sources errors pertaining to those instruments are mixed to give a different result. When very good accuracy is desired data should be collected from high end equipments. Even if one adds the data from a comparbly less accurate instrument to the data of many accurate instrument is then the result obtained finally will not reliable enough. Global Positioning System (GPS) is available with mm accuracy to meters accuracy. Also surveying with suitable map projection and datum is important; otherwise the result will be wrong. Survey instruments posses some inaccuracy and when a person is using it he should have the accuracy over the instrumental like rain gauge, water level recorders which give attributed data which should be checked for instrumental inaccuracies and method of data collection before using it. Likewise satellite and aerial sensors miss some data and some error are inherent. Such errors should be removed before moving ahead. Errors are resulted due to map reduction, enlargement, data editing and computation method adopted. At the same mapping instruments also possess inaccuracies. When after certain interval a person is going for survey, the registered object may have been changed. Many new features might have come up. Hence the established control points are changed and if the wrongly placed control points are taken for survey then the result itself will be wrong. If proper resolution is not available in the instruments, then the instruments may not capture the minute details which are required.

5.3 ERRORS INTRODUCED DUE TO DATA PROCESSING IN GIS

Errors tends to take place, while making data entry. If the distorted map is used as a backdrop for dizitisation, the features in the map will not be on their actual position, thus resulting in wrong results at later processing. Manual digitisation or automatic vectorisation leads to many errors. Also conversion form raster to vector and vector to raster introduce errors. Manual digitisation is a time consuming and tiresome job. A person may be in a better position to work for two continuously. When he continues his work for many hours he may lack concentration due to which some digitisation errors may be left with. While entering attribute data instead of entering, 100 if 500 is entered, then the value 500 is taken for interpolation thus giving wrong result.

Digitisation is the major work in any GIS project. Conversion of paper map into scanned documents is an easy job. But conversion of raster data into vector data is not an easy job. Many softwares are available for vectorisation. AutoCAD and Microstation are the most widely used CAD softwares for digitization besides most of the GIS softwares posses their own commands for vectorisation. During editing a lot of errors occur. It is important to remove all the errors before analysis. Also data updating is important to cope up with the developments in various sectors. Road maps, parcel maps, water distribution maps, sewer maps, land parcel map, landuse landcover map *etc* are need to be updated frequently. In this case digitization of the new maps with the existing maps is important.

Also errors occur in scanning and tracing. The tracing algorithm adopted in the software may act differently when raster lines meet or intersect, or they are too close, too wide, too thin and broken. Due to this the error like collapsed line, misshaped lines and extra lines will appear.

Due to georeferencing error may occur. If one compare the original map and georeferenced digitised map, he may observe the difference. If he is able to measure the same distance between two locations in both the map, then the digitized and georeferenced map is correct. If accurate Ground Control Points (GCPs) are given to georeference the map, then the result is accurate.

Duplicate lines may occur during manual digitization or automatic digitisation. Each line or polygon is to be digitised only once. But due to manual error or automatic digitisation in complex map draw the lines twice(called slivers). Without zooming it is not possible to find the slivers. If topology is built the slivers will have polygons without labels and this is the error.

Pseudonode is an unnecessary node existing in an arc. Even sometimes psudonodes are needed. In an isolated polygon, a pseudonode will exist. Arc directions are also important in many applications. In one way road there will be transport from one node to another node(from node to-node). Also streams are starting from high elevation and reach lower elevation.

Few softwares are based on topology while others are not

AutoCAD Map, ArcInfo, Microstation Geographics, ILWIS, Geomatica, Geomedia Professional are based, on topology. When a data is created, they are edited topologically. Topology errors are identified and cleaned automatically. But if data is created in a non-topological software, it is difficult to identify these errors. The data can be imported into topology based software and they can be thus edited. Softwares like ArcView does not possess topology and cannot handle topological error.

Fig 5.1 to 5.6 shows the errors due to digitisation. If an overshoot and undershoot occurs then a tolerance may be set to remove errors.

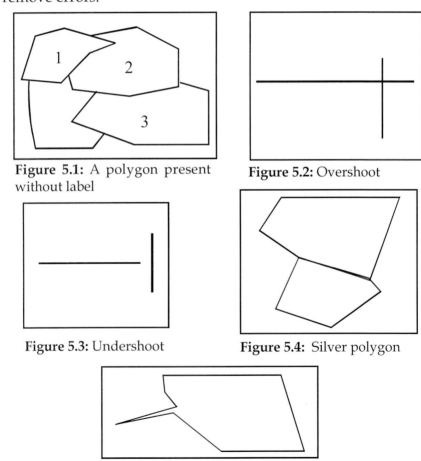

Figure 5.1: A polygon present without label

Figure 5.2: Overshoot

Figure 5.3: Undershoot

Figure 5.4: Silver polygon

Figure 5.5: Spike

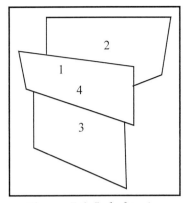 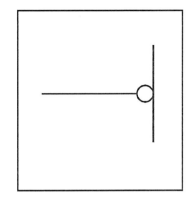

Figure 5.6: Label twice　　　　**Figure 5.7:** Tolerance to remove

To have good result, computers with sufficient memory and precision is required. Storage medium may also have errors sometimes. Converting raster to vector and vector to raster may result in errors. Many interpolation techniques are available and proper interpolation techniques should be used for a particular work. The user should have knowledge on the interpolation techniques and he should use the right choice. This is because of the varying nature of data over ground. Plotter used may not plot the expected result correctly due to the incompatibility with the software, computer or the plotter.

5.4 ERRORS IN METHODS

Good amount of expertise is required to collect and store the valuable data. Samples collected from proper locations are always needed to construct a real world model. Insufficient observations density of points may give wrong results. The features should be registered properly. Uncertain border between areas should be handled with due care.

5.5 METHODS TO CORRECT EXISTING ERRORS

In the above paragraphs various sources of errors are identified and it is the role of the GIS user to avoid and correct such existing errors. Frequently various process should be checked. If the errors are checked and removed at every stage, then the result is good and time is also saved. If the project is not checked for errors and if the errors are noticed at the end of the project then it would be difficult to remove those errors. It is like starting the work again from the scratch. The digitised map may be taken as a printout and checked for digitisation errors. Data from high accurate sources and low accurate source should not be mixed. Proper data processing, data classification should be done.

Firstly, during this process, the software will identify each node, line and polygon. Next, the type of errors will be identified. The errors are removed. Many types of topological errors exits. One type of error may be removed once or all the errors may be removed at one. But even the users can interact during the removal of errors. After all the errors are removed , topology is rebuilt.

Removal of errors involve to set tolerances. A specified dangle length will be given and any gap in line or polygon less than the dangle length is closed. Tolerance may also be given in numbers instead of drawing a dangling line. In case of duplicate lines, fuzzy tolerance may be set to remove errors. The tolerance unit set in digitize will be in inches or it may be given in meters or feet. More tolerance will distort map features. Due care should be taken while setting the tolerance. The errors may also be identified by zooming in to correct it manually.

A node may be inserted by splitting a line. Unclosed polygon may have two labels and this error may be removed.

Data Accuracy

Quality of a data is based on the fact that how much complete consistent accurate, precised, timely and above all effective data is available.

Accuracy

This expresses as true value as what actually exists. Otherwise accuracy may be defined as the deviation between measured value and the true value. Numbers can be mistyped, changing the location name inadvertently. So a GIS user should know the nature of data, nature of problem, its coverage, its influence *etc.*

Precision

Precision is the depiction of details or significant digits in measurement. Objects with millimeter distance is precise than with centimeter. If anyone is interested to locate a point based on the data and he is having problems in identifying the point then the point is not precise. Precision and accuracy are closely associated. Generalization of digitized lines give imprecision. Fig 5.8 shows the relation between accuracy and precision.

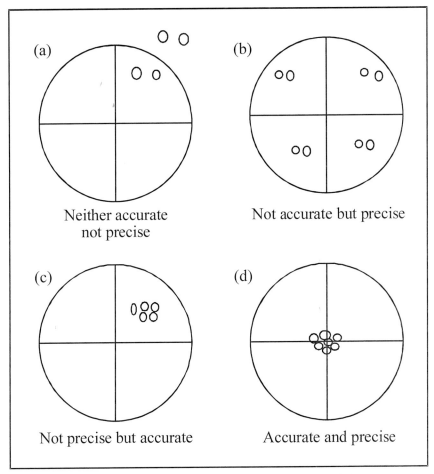

Figure 5.8: Relation between accuracy and precision

Edge matching

Many maps are used in a single project often. Maps are digitized individually and removed for errors. Then the two maps to be combined will be kept side by side and lines passing between the two maps are identified and joined using edge matching command. After matching the maps, the borders will be merged and two maps now become a single map (Fig. 5.9).

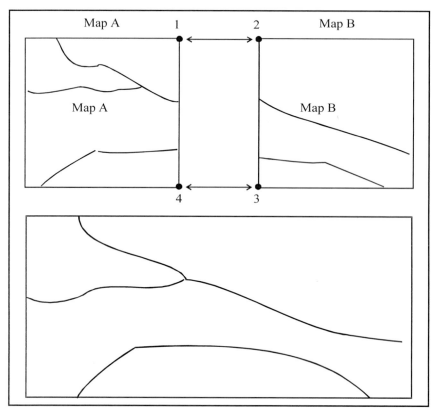

Figure 5.9: Single map obtained by edge matching technique.

5.6 GIS OUTPUT

A map is a communication tool of spatial information. A map should consists of title, body, scale, extent, north arrow, neat line, extent of map coverage. Map body is an important component as it possess major information of an area. Other elements of the map are used to supplement the map feature.

Errors in GIS and Data Output

Title tells about the map subject, scale tells about the distance of the area covered in ground *etc.*, Map making has now become easy due to widely available GIS packages. Formerly separate cartographic packages were available for producing maps. But it is possible to change scale of the map, legend of map, title *etc.* very easily. Maps should be designed with a clear idea in mind. A well designed map will convey more information, where as on the other hand a poorly designed map will only confuse the reader.

Results from GIS are important as the analyzed information should be communicated to the end user. The results are useful for decision making.

The output is derived from the given data, method of input, manipulation, types of analysis are carried out. Hence the output is based on these processes. If an error occurs in any one process or combination of processes, then the result thus obtained will not be much useful. A map is produced on the basis of cartography which is the art and science of producing maps.

A map communicate spatial information to the users. Hence a proper map design is very important. The information available in the maps are point features, line features, polygon features and annotation about them . The map also should include scale, study area extent (latitude, longitude), an insert map showing location, legend, north arrow, title, reference grid, meta data like projections used, accuracy of map. These should also follow the cartographic symbols.

A map should not contain any other information which is not relevant to the scope of the work undertaken, otherwise this will reduce the importance of the project taken. A map should not contain too much information or too little information either. Map generalization is important. If one intend's to show the road network of a region, he may avoid very small roads, the reason is, the presence of small roads will occur, throughout the map and thus the readability of map will be lost. As the real world objects are shown as point, line or polygon, they should differ among each by their size, shape, density, texture, orientation

106   *GIS: Fundamentals, Applications & Implementations*

and colour. Also mapping agencies follow cartography to produce maps and such map design is important for the GIS users. Rivers are shown by blue colour. Black and white maps use proper symbols and shades and they are cheaper but the colour maps should be represented very carefully as they give more information than the black and white maps. Fig 6.26 in Chapter 6 shows the cartographic symbolism.

Also different symbols are used for different utilities. Telephone lines, power lines, rail roads, lighthouses *etc* are shown conventionally with particular symbols. If the symbols are interchanged, then the entire meaning and purpose of preparing and presenting the information by means of maps is lost.

Data in a map is classified in different ways. For presenting population data chloropleth maps are widely used. Chloropleth maps shows equal distribution of information. In monochrome maps, high shaded region shows high value and less shaded region shows less value. Solid white and solid blacks are avoided in maps conventionally. Higher value regions are shown with more closed parallel lines and less value with less parallel lines.

Conventionally data in a map can be classified into fine classes. More number of classes will lead to confusion. Hence they shall be regrouped into five classe.

Meta data information should be present in maps. This will help the user to know the method of survey or observation, accuracy of the instruments, method of interpretations, year of survey. This is just like the warning smoking is injurious to health, printed on cigarette packets.

If temporal variations are considerably larger, then animation can be used to see the changes. Some of the process like urban growth, population explosion, pollution in the atmosphere, water pollution can be mapped for different stages. If they are animated, then the changes can been seen very easily.

Non cartographic outputs like tables, graphs and text information are also useful. Per capita income of various states shown in a pie chart will give good idea of exploring the data immediately. The shortest path for disposing the solid waste is given as text information so that the truck driver can use this information to carry out his work properly and well on time. New table from an analysis may be used to know the statistics of the information. Also linking a map or image or table or document with map is possible in GIS. The photograph of the study area, population details of the area can be shown in a single map for better visualization and comparison of data for decision making. Some GIS softwares also support audio-video visuals thus making spatial multimedia also possible.

Monitor is one type of an output device available to a GIS operator. To show it to others, the maps are printed using various high end colour printers and plotters available now-a-days.

Virtual Reality GIS is also available now-a-days. Here GIS is linked with VR softwares and one can virtually fly through a hilly terrain. For producing a map in GIS, the analyst need not be a professional cartographer. GIS software posses their own symbols for representing all the features and hence the role of a GIS user is to just add to it. Automation is done for adding legends, scale and keys. The data available in a map can be symbolized differently.

Unique value map: This is used when different colours are to be used to symbolise different types of things in a single map. For example 4 types of soils are represented in 4 different colours. Each state in a country will be shown in a unique colours.

Graduated colour map: This is used when there is continous increase in the value. For example temperature of a city. For ranking of 1, 2 and 3 this method is good. For data with numerical progression is like measurements, rates, percentages these maps are of great help.

Graduated symbol map

These maps are similar to graduated colour maps but the difference is instead of using colour, symbol with different sizes

108   *GIS: Fundamentals, Applications & Implementations*

are used or lines with different width are used. Road may be classified according to traffic volume or cities with different population. Hence the relation smallest to greatest is found visually.

Dot density map

Dots are used inside the polygon to represent an attribute value. In a map, if a point is represented for 100 people, to represent 2000 people in a polygon, 20 dots will be presented in the map. Graduated colour map is used to represent the population difference. If two states possess the same amount of population, same colour will be used to represent the states in the map. But in dot density map, even though the population of the two states is same, states with small area will have closer points and states with larger area will posses points at some distance. Hence the density of population is well represented in dot density map.

Chart map

Here bar chart or pie charts are used for the points or polygons. For example various income groups of various cities may be represented in city maps. When one compare the pie chart for these cities, different income groups can be clearly seen.

5.7 Classification method for graduated colour map of graduated symbol map

Following are the classification methods used in GIS

- Natural break
- Quantile
- Equal area
- Equal interval
- Standard deviation

Besides the above mentioned classifications, it is possible for a user to give his own classifications also.

Natural breaks: This method groups the data when there is a major break in the data flow. Hence the variation of data in the same class will be minimum. But variation between groups are sharp. For presenting population data this is most suitable.

Quintile: For each class same number of features are given. This method is suitable when data is linearly distributed.

Equal area: This method is used to classify polygons by finding break points in the attribute. So total area in each class is approximately same. This is like quantile classification but the difference is each feature will be given a weight in the classification equal to its area. Hence smaller states will not show variations among them when the map consists of bigger states.

Equal interval: This method divides the data into equal sized sub-ranges. Here if value is ranging from 1 to 300, then if we would like to divide this into 3 classes each class will have a value of 100 *i.e.* 1-100, 101-200, 201-300. This method is more when one is interested to show even the smallest variation among the classifications.

Standard deviation: This shows the extent to which an attributed value is differing from mean value. The classification from mean *i.e.* 1, <3 or >3 standard deviation.

Colour Models

Computer screen uses Cathode Ray Tube (CRT) to display colour. CRT consists of red, green and blue colour guns with each of 8 bit level. Due to the combination of the three colour at 8 bit level, 16.8 million colours are generated. Eight bit can produce 256 levels of colours (0-255 level). Hence each 8 bit RGB (256x256x256) produces 16.8 million colours. Red, Green and Blue (RGB) are the primary colours, while Cyan, Magenta, Yellow and Black (CMYK) are sub-tractine colours and are thus used in printing because they print what actually appears on the computer screen.

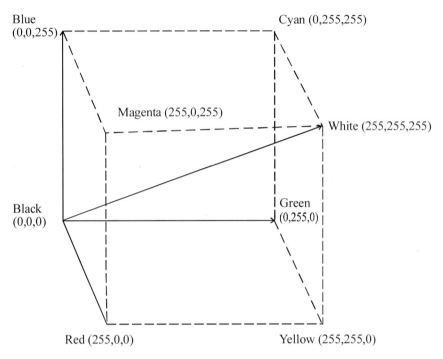

Figure 5.10: Red, Green, Blue, (RGB) and Cyan, Magenta, Yellow, Black (CMYK) Colour Model.

Advanced GIS Applications 6
— Case Studies

6.1 INTRODUCTION

Besides simple analysis, GIS is useful to do advanced analysis like network analysis, spatial analysis, 3D analysis and modeling. Different softwares are available to perform such analysis and the above mentioned analysis is done in separate modules of the softwares.

6.2 NETWORK ANALYSIS

Network consists of well connected linear features. Roads, railway line, water distribution system, sewer line, streams are some of the examples of networks. These networks consists of nodes and arcs with designated directions and connection with other linear features. Networks are topology based with attributes for the flow of objects like traffic. Road network will be considered for explaining network analysis.

Road map is digitised with nodes and arcs. Signals, accidents are represented as points. Road is represented as line and two or more roads intersect at nodes. Attributes are added to the nodes and lines. Dynamic segmentation model is used for network analysis. It is built on lines of network. In road network analysis, attributes for travel time, one way street, two way street, right turn, impedance are added to the nodes and arcs. A link is the line running between two nodes. Link impedance is the cost of passing through the link. Travel time is different in links in cities due to varying traffic at different locations. Time taken to pass through a known length of link may give the cost but the travel time in a link is not uniform because of varying speed designated in the link as 20 km per hour and 30 km per hour. Hence it is important to consider the length of link and also the travel time. If the speed limit is 20 miles per hour for a length of 2 miles, then the travel time is 6 minutes (2/20x60 minutes). Also tuning in the network is considered. A road may have a maximum of 12 possible turns excluding U turns. There are right turn, left turn, straight way for each road. Also U turn is possible. This turning takes different time depending on the signal. Straight way takes less time but left and right turn may require some more time. If signals is available then signal time should also be taken into account. Fig 6.1 show the possible turns in a road.

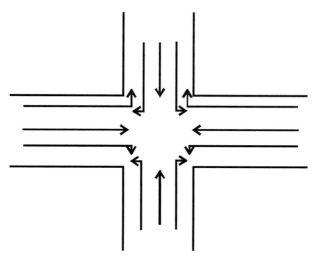

Figure 6.1: Possible turns in a road

One way, closed roads, two ways, overpass, underpass, are represented in road network for effective analysis. If two road intersects, a node is created. But when an overpass passing above a road, no node is present. In the same way, when an underpass passing below a road do not have a node. Some of the analysis have been already dealt in previous chapters. In this chapter spatial analysis, network analysis have been dealt with case studies.

Topology

They are node topology, chain topology. When a road network analysis is done, the software is not able to find the route. Hence the user has to give from node, to node, overpass, underpass, road cross, left turn, right turn *etc.*, Normal digitization produce spaghetti model without topology. It is important to build topology to have the relationship of objects. The relationship is among points, point to line, point to polygon, line to line, line to point, line to polygon, polygon to polygon, polygon to point and polygon to line.

Topology is useful to find shortest path from one location to other. Occurrence of a lake within 10 km radius of a dense forest for serving water to wildlife.

Contiguity Function

Common relation among map features may be found based on the contiguity operator. In a landuse map, if one wants to find the association of river intersecting a lake and the river is 500 m away from road. The areas which posses lake intersecting river and the river from 500 m are selected. This may be used for any required purpose.

Proximity Function

This is determined by creating a buffer zone or Thiessen polygon or by drive time analysis.

Industries located within 1 km radius may pollute the river. The query may be to shift the industries from 1km buffer zone. Industries which are proximal within 1km is identified and they may shifted to some other areas where rivers are not within 1km radius. Thiessen polygons may be constructed to find the area with same amount of rainfall or for a business center, the area served within 5 km radius may be found. Banks or any marketing centers may set up based on the service area identified based on Thiessen polygon. A person may be interested to go to a hotel within 30 minutes drive time and thus hotels can be founded by analysis.

Connectivity

Connection between features like a state highway with national highway. A lower stream connected with higher order stream

Contiguity

This is the degree of connectivity. In a map lakes and forest occurs. One may impose a query like find the lakes which are connected with bitumen road. Then only maps which are linked with bitumen road will be displayed.

Adjacency

Features close to each other, Find a five star hotel within 5 km radius of airport. Hence closeness of one object to other is established. Four locations A,B,C and D are available. The closeness among points may be identified.

Time and GIS data

Development in urban are may be found if maps for two different period available. If 1990 and 2005 maps are available for a city area, just subtracting 1990 map from 2005 map will show the new developments over 15 years.

Also if the landuse maps for two different periods are available, the subtracting one with other show the change in landuse pattern and such maps will be useful for urban planning.

6.3 SPATIAL ANALYSIS

Raster simple overlay

Maps consists of numbers, hence tow maps are added, divided, subtracted and multiplied. Fig 6.2 shows the different simple overlay.

Weighted overlay is required in many cases. Landslides may occur in a location due to many causes. Rainfall, slope, drainage, soil type, landuse are some of the factors. But all are not equally responsible. Rainfall is the major criteria for landslide and after that slope may be important. Hence based on the influence of a theme the theme is ranked. Fig 6.3 shows the weighted overlay.

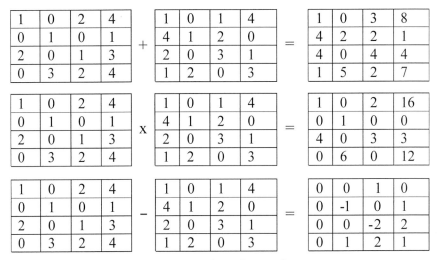

Figure 6.2: Simple overlay

Spread Function

If DEM and flood map is available for a dam area, the volume of water available and area inundated by flood is found.

Seek Function

If DEM map is available for terrain, then drainage is created. Also Strahler's stream orders are found after the drainage is created.

6.4 TERRAIN MAPPING

Land surface is mapped for various applications. GIS is very much useful tool in developing contour. Maps like Slope, Aspect are derived from this. These maps are used in various applications. Terrain mapping uses either vector or raster data. z.value is the elevation.

DEM (Digital Elevation Model)

DEM in obtained from contours. In contours only at the lines we have the elevation. But after interpolation each and every location consist of an elevation. USGS supplies DEM data but this DEM is different from the DEM obtained using interpolation in GIS software.

116   *GIS: Fundamentals, Applications & Implementations*

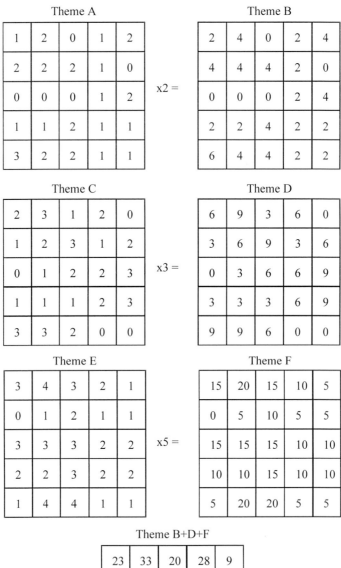

Figure 6.3: Weighted overlay

TIN (Triangulated Irregular Network)

Here the TIN is created using the z data by non overlapping triangles. TIN consists of irregular triangles. Sources of TIN are from surveyed z data, DEM, contours. The triangles are connected using the algorithm Delaunary triangulation (Fig 6.4). The rule is to connect all nodes to nearest neighbour and the traiangles are equiangular or compact. Along the border the triangles are distorted. To avoid this triangles may be drawn outside the area of interest also and clip the study area from this.

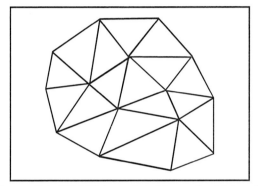

Figure 6.4: TIN (Triangulated Irregular Network)

Thiessen polygons

Voronoi polygons and Dirichlet polygons are the other names for Thiessen polygons. An area is divided into many polygons consisting of a single point. Any point inside a polygon is closest to the centre point of the polygon than any other center point in other polygons. Thiessen polygons are also used for interpolation. Service area covered by a store may be identified.

Interpolation

It is assumed that no significant error exists in the sampling, the points are well distributed and the pattern changes in a regular manner. Thiessen polygon is a type of interpolation. All changes are taken into account. This is like stepped terrace. Changes are taking place at the edge of the polygons.

Spline

Spline fits a curved surface to points. The curve is smooth, going through all the points.

IDW (Inverse Distance Weighting)

The interpolated contour is the weighted mean of the observed values.

IDW Inverse Distance Weighting

It is assumed that the points around a location are similar to the point than the points away from the point. The value interpolated is by considering the surrounding points. The weights are the square of inverse distance. IDW used a power of 1 or more divided by the distance as the weight. If higher the power the influence of distant points are less, number of points to be considered for interpolation may increased or decreased.

Trend surface

This is just like fitting a line through scattered points. But this is not widely used for interpolation . The fitting is based on linear regression or least squares.

Plane or polynomial or trigonometric surface are used as trend surface to fit.

Kriging

This is like IDW but the weightage is varying from the distance. Points which are closer have less variance and farther have more variance. Variance reaches maximum at some point of time and after which there is no change. Then interpolation is carried out based on the variogram developed. Variogram is a graph which relates how similar points are closer and that are compared with points farther. For kriging more number of points (atleast 50) are required even it is possible to use with less number of points.

If point data is available, then the data is interpolated with any of the above techniques. Fig 6.5 shows the point map with

Advanced GIS Applications 119

attributes. Fig 6.6 show the DEM created and Fig 6.7 show the contour map. Fig 6.8 show the slope map and 6.9 show the aspect map. Hill shade map is shown in Fig 6.10. Hence if spot heights are known it is possible to derive the above mentioned maps.

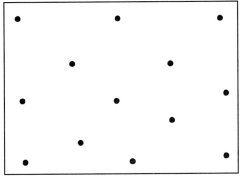

Figure 6.5: Point locations

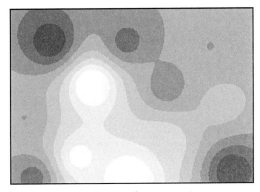

Figure 6.6: DEM (See colour version on page 185)

Figure 6.7: Contour Map

120   *GIS: Fundamentals, Applications & Implementations*

Figure 6.8: Slope map

Figure 6.9: Map Aspect (See colour version on page 185)

Figure 6.10: Hill shade map

Fig 6.11 shows the Boolean overlay. They are AND, OR, NOT and XOR. These Boolean or logical overlay are used to extract information based on the overlay.

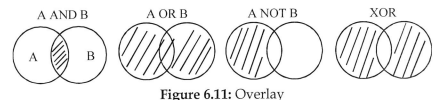

Figure 6.11: Overlay

Fig 6.12 shows the intersect overlay. Fig 6.13 show the point in polygon overlay, Fig 6.14 show the line in polygon overlay and Fig 6.15 show the union overlay.

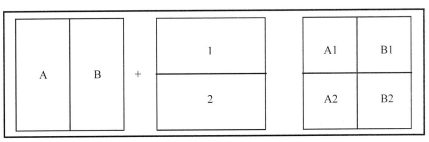

Figure 6.12: Intersect overlay (Polygon on Polygon overlay)

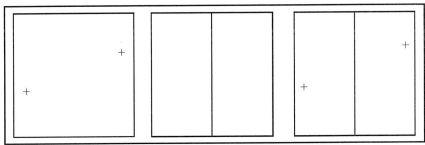

Figure 6.13: Point in polygon overlay

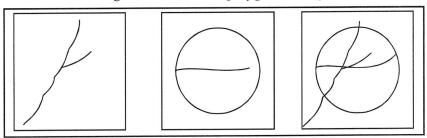

Figure 6.14: Line in polygon overlay

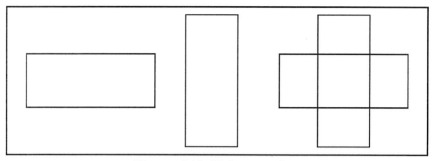

Figure 6.15: Union overlay

Catchment

Catchment is the term used in hydrology, *i.e.* the concentration and movement which water into a single point. GIS software automatically derive catchment area but doing the same manually require more time. Catchment is derived from contour and from DEM. Unwanted peaks and pits are removed. Aspects and curvatures are derived. Then flow of cell direction is obtained (Fig 6.16). Flow of water is identified from this, hence drainage is derived.

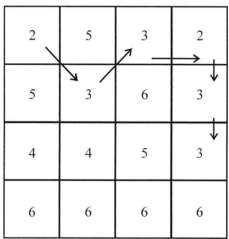

Figure 6.16: Shortest route in raster data

Flow directions in a basin is in four directions or in eight directions. They are represented as in Fig 6.17 and 6.18. Raster models are good in representing such flow directions than vectors.

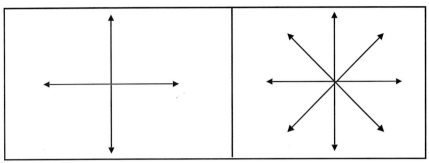

Figure 6.17: Flow directions (Rooks Move)

Figure 6.18: Flow direction (Queens Move)

From contour map, slope is created and calculated as shown in Fig. 6.19. Aspect is shown in Fig 6.20. Aspect is the direction along which water will move.

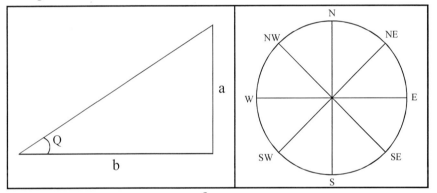

Figure 6.19: Slope percentage is $\frac{a}{b} \times 100$

Figure 6.20: Aspect

6.5 CASE STUDIES

6.5.1 GIS Based Water Distribution System for State Insurance Housing Society, Coimbatore

6.5.1.1 Introduction

Geographic Information System (GIS) is very much useful in storing, maintaining, updating and analyzing various utilities like water distribution system, sewer system, power lines, telecom lines, transportation network *etc.*, GIS is a powerful tool for query and visualization. Once the data for utilities are entered, updating of new data is quite easy. If any problem occurs at a particular location in the network, it is easy to identify in GIS using the analysis modules available. Diameter of the pipe, pipe material and other related information could be obtained through such simple queries and the utility engineer could know the nature of repair work to be done and he will be able to go the field with all preparations before hand. Also the location of pipes is shown in the GIS map and digging is made at the right place.

6.5.1.2 Study Area

The study area State Insurance Housing Colony (SIHS) colony belongs to Coimbatore Corporation of Tamil Nadu State. For Coimbatore corporation, water is supplied from two different schemes namely Siruvani (West Zone) scheme and Pillur (East Zone) scheme. A total water quantity of 838 lakh litres per day, has been supplied to the Coimbatore city. For the above mentioned schemes there are already 25 service reservoirs existing with pipe distribution length of 351 km.

Due to uncontrolled growth of population, many new housings and residential zones are developed. For the growing population, the basic need 'the water' is supplied at the rate of 110 liters/day/person. To meet this consumption level, Coimbatore corporation implemented a new scheme of water supply system for the Coimbatore city with two major zones *i.e.*, the east zone with 20 service reservoirs, water supplied through Pillur scheme, with a new distribution pipe length of

95 km, and the west zone, water supplied through Siruvani scheme with 22 service reservoirs, water supplied through Siruvani scheme with a new distribution pipe length of 170 km.

Study area SIHS colony comes under the east zone of Water supply system.

SIHS Colony is located within the corporation limit in the eastern part near the Singanallur railway station. It comes under the Pillur water supply scheme. Through this scheme water is supplied to a population of about 8485 inhabitants. Water is supplied at the rate of 110 lit/cap/day. The total capacity of the service reservoir is 3 lakh litres. The ground level at that point is 400.000 metres above mean sea level and the top level of the reservoir is 415.650 metres. The height of the tank above the G.L. is 12.00 metres. The total length of pipe through which water is distributed is 8300 metres. Total quantity of water distributed through the tank per day is around 9.00 lakh litres. The tank is located at the SIHS Colony itself and it distributes water only to that colony. The total estimated cost of this project was Rs. 46.62 lakhs.

6.5.1.3 Data Collected

The spatial data collected were maps with details about different types of pipes *i.e.*, main, sub-main, OSM, branches *etc*, their names, street names, different types of valves and their locations, direction and quantity of flow of water, pipe / street length, contour levels *etc*.

The non-spatial attributed data collected were population details, litre per capita per day details, required flow in litres per minute for each and every street and pipe systems, quantity served through the pipes, reservoir details, possible losses, *etc*.

6.5.1.4 Methodology

Map for the SIHS colony was obtained, scanned and saved in TIFF format. This map was inserted in the AutoCad Map 2000 and 11 layers were digitized. The map was cleaned for various errors and topology was built.

126   *GIS: Fundamentals, Applications & Implementations*

Then to have a better visualization and data analysis, the map was brought into Arc View. After starting ArcView, Cad reader extension was loaded. The layers created in .dwg format were added by add the theme buttons. All the eleven layers were available in a single theme. Hence to create individual themes, from theme properties, drawing was selected. All the eleven layers were available in the layer menu of theme properties. Legend editor was opened to have unique value type as legend type and layer as the value field. During this conversion process, all the attached attributed data in AutoCad was lost. Entity, layer, elevation, thickness and colour were the field type available and the data was no longer relevant. But layer and colour fields were useful to give attribute data once again in the Arc View table.

The layers were digitized in AutoCad Map and brought into ArcView are given below. All the eleven layers in .dwg format were converted into shape files.

— Water Tank
— Sub-main pipe
— Street
— Sluice_80-valve
— Sluice_100-valve
— Scour valve
— Opposite side main
— Main pipe
— Flow
— Contours in meters
— Branch Pipe

Data attached for service reservoir

The data given in the table for the theme water tank are the names of the areas served, total cost of the project, population served, tank capacity, place of location of tank, ground level, top level of the tank, maximum water level, lowest water level, height of the tank above ground level, total length of distribution system and quantity of water supplied per day.

Data attached for branch pipe

The data provided in the table for the theme branch pipe are branch number, starting distance of the pipe, ending distance of the pipe, total length of the pipe, quantity of flow through the pipe, size and material of the pipe, grade over, velocity of water flow in the pipe, head loss due to friction in the pipe, head loss due to other reasons, total head loss in the pipe, head at the inlet of the pipe, head at the outlet of the pipe, ground level at that location, residual head at the pipe outlet, connections to other pipe, quantity distributed through the main pipe, quantity distributed through the sub-branch, total quantity of water distributed and name of the area / street served.

Also data for themes main pipe, opposite side main, sub main were same as that of the data given for the theme branch pipe. Fig 6.21 shows the water distribution network for SIHS colony.

Data attached for streets

The data given for the theme street were name of the street/ area, name of the pipe(s) serving water, length of pipe, population served, required quantity of water, peak demand quantity

6.5.1.5 Querying The Maps

The query by information tool provides information just by mouse click. Map Query Builder was also used to ask arithmetic and boolean queries.

Using query builder, a simple query [Layer] = "STREET") was executed. This query highlights all the streets in the study area and shown in Fig.6.22 . Like wise queries were performed to identify the required features for the remaining themes.

Also the following queries were performed:
- (i)   Show all the branch pipes.
- (ii) Show all the pipes of length greater than 50 metre.
- (iii) Show all the pipes with a head loss more than 0.5 meter.

128   GIS: Fundamentals, Applications & Implementations

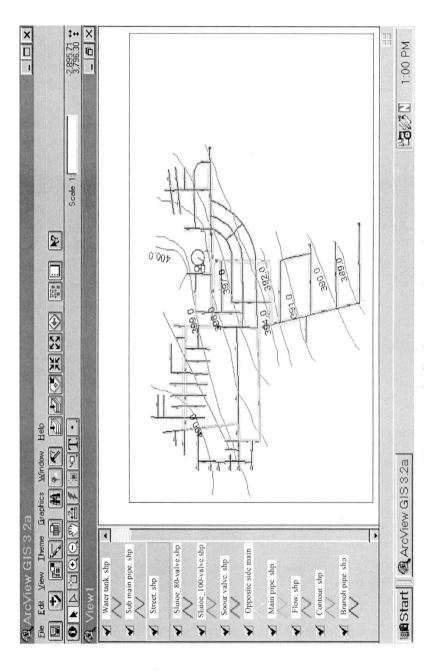

Figure 6.21: Water Distribution Network Map for SIHS Colony

Advanced GIS Applications 129

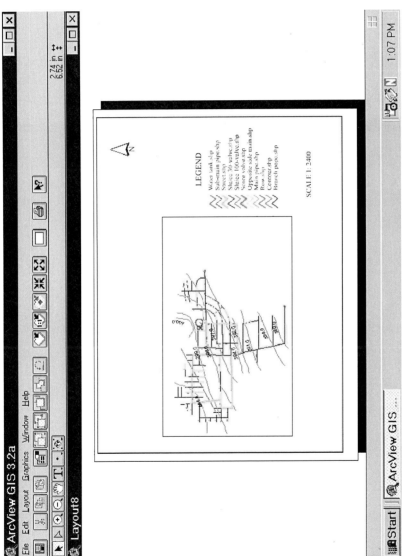

Figure 6.22: Query By Map Builder To Select Street Theme

130   *GIS: Fundamentals, Applications & Implementations*

As the data for various themes are available in the utility information system, breakage or if any other problem is reported in a particular location, the service engineer can go to the field with due preparation. This study indicates that GIS based information system for utilities are very much useful in the present day for better management.

6.5.2 Route Optimization For Solid Waste Disposal For Coimbatore Municipal Corporation

6.5.2.1 Introduction

Humans and animals have used the resources of the earth to support their life and used materials were disposed as wastes. In early times the disposal of human and other waste did not pose a significant problem, as the population was small. But due to the population explosion and developments in various sectors, a lot of waste is generated. In cities, solid waste disposal is a major problem. Coimbatore is the third largest city in Tamil Nadu with a population of more than 13 lakhs, including a floating population of around 1.5 lakhs. With the rapid increase in population, nature is moving towards a total breakdown and ecological imbalance of water and air pollution have been attributed to improper management of solid wastes.

Solid waste management is one of the most important areas where the problem arises from time to time. Municipal bodies are unable to provide a 100% efficient system for solid waste management. Solid waste management frequently suffers more than other services. The provision of collection and disposal services for municipal corporation do not get higher priority. The real problems are mainly of organization, management and planning.

Urban development has brought forth several maladies and suffering to human kind, besides bringing economic and cultural development in its fold. Due to pressure of urbanization most of the cities are growing and some times they develop beyond the planned limits. Generally the unplanned area of the city contains a quarter of the total population where the spatial information is missing. Solid waste disposal is a major problem for the Coimbatore Municipal Corporation. An attempt has been made to optimize the route and also finding a favorable site for disposal of solid wastes using GIS and GPS for Coimbatore Corporation.

6.5.2.2 Study Area – Coimbatore Municipal Corporation

Coimbatore popularly known as "Manchester of South India" is situated in the western part of Tamil Nadu. Coimbatore is well known for its textile industries, because of its proximity to the hills of the Western Ghats, Coimbatore enjoys an excellent climate through out the year. Coimbatore city was constituted as a municipality in 1866, with a population of 24,000 covering an extent of 105.56 sq.km. Coimbatore city has been elevated as a municipal corporation from 1981. The study area has bounded between 10° 58' N to 11° 30' N and 76° 55'E to 77° 30' E. At present the population of the city is approximately 13 lakhs, including a floating population of around 1.5 lakhs with an area of 105.56 sq.km. The city has been divided into four zones namely, North, South, East, and West. Where by each consists of 18 wards, total of 72 wards.

6.5.2.3 Objective

Objectives of this study are:

To optimize the route for solid waste disposal in Coimbatore Corporation using GIS

To create GIS based database for solid waste disposal

6.5.2.4 Methodology

The details of the container location for various zones have been collected. Various attribute data like container location, ward number and zone, container number is created in the arc pad software before going to the field. After that the data's are stored into file as shape (.shp) file.

Leica GPS GS5 has been used to locate container bins and dumping yard. This GPS consists of an antenna, connecting cables, GPS kit bag, HP iPAQ Pocket PC, two batteries, Arc Pad software and interface to transfer the data from GPS to computer.

The GPS instrument is taken to the field and map projection and scale was set in the software. GPS was activated in front of a container. It started receiving signal from the satellite. After few seconds or minutes the latitude, longitude and altitude of this point were displayed and the data is stored. Likewise latitude, longitude and altitude for the containers were identified. Then the GPS was brought to the lab and the data was transferred into computer using the interface Microsoft Active Sync. Map of Coimbatore Corporation consists of roads and facilities was added in Arc View and the containers location were added.

The vehicles move along various streets and collected solid waste from several collection points and when full, proceeded towords the transfer station/processing/disposal site. Presently the routes of the movements of vehicles are arbitrarily fixed, usually by the supervisor on the basis of his experience and convenience. These routes are not necessarily the best routes and certainly not optimum thus results in under utilization of vehicles and increased cost of transportation.

Fig 6.23 shows the flow chart of the activity of the study undertaken. The existing condition of solid waste disposal was studied. The base map of Coimbatore Corporation was scanned and inserted into AutoCAD and digitized for various layers like roads, streets *etc.,* This map was imported into ArcView and converted into shape file. This shape file was transferred into Data Automation Kit. Various digitization errors were removed and topology was built using the same software. Then this file was brought again into ArcView and attributed data were added. GPS locations with their details were also added into ArcView. Then the shortest path between the containers and dumping location was found using the network module for north zone which is shown in Fig.6.24 The comparison of the actual route and the route derived from this study using GIS is shown in (Fig.6.25). Likewise the shortest path analysis was done for south, east and west zones and the difference in kilometer per trip for each zone is presented in Table 6.1.

Advanced GIS Applications

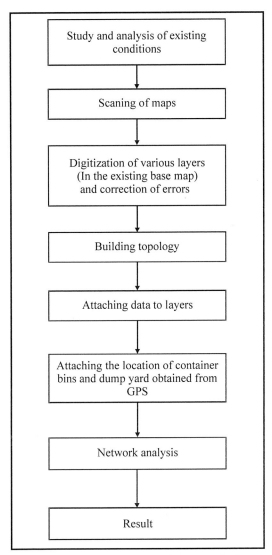

Figure 6.23: Flow chart of the work done

Table 6.1: Overall difference in km from container bin to dumping site for each zone

	North Zone	South Zone	East Zone	West Zone
Distance reduced in km per trip	29.37	7.93	32.21	19.6

134 GIS: Fundamentals, Applications & Implementations

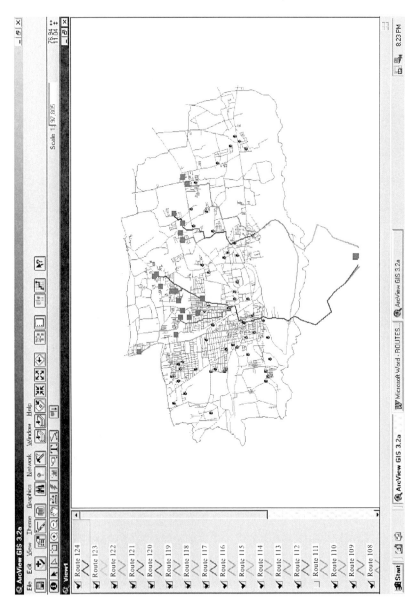

Figure 6.24: Shortest path for the North Zone from container bin to dumping zone

Advanced GIS Applications 135

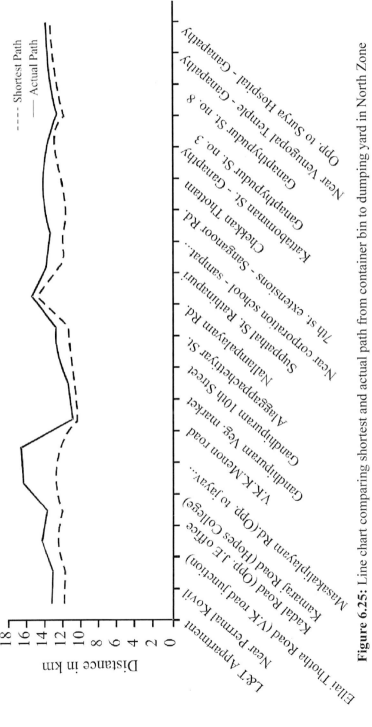

Figure 6.25: Line chart comparing shortest and actual path from container bin to dumping yard in North Zone

136   *GIS: Fundamentals, Applications & Implementations*

It is found that if the route selected by the present study is adopted then 4.08 km can be saved there by diesel consumption of about Rs.15/- can be reduced per trip per truck.

6.5.3 Site Selection For Solid Waste Disposal

6.5.3.1 Introduction

Due to the rapid industrialization and increase in population large quantities of wastes are being generated in different forms. At present Coimbatore Corporation face one of the most critical problems, which relates. The solid waste produced around 638 tonnes per day. An attempt has been made to find a new sites for dumping of solid waste disposal. Due to the opposition of the publics of Vellalore, dumping site has become controversial. In this situation, an attempt has been made to find a new site for dumping of solid waste disposal.

6.5.3.2 Methodology

For selection of new sites the data like landuse map and soil map has been collected. The methodology is shown in Fig 6.26.

For selection of new site the data like land use map and soil map of Coimbatore taluk, which is for five blocks (Thondamuthur block, Madukarai block, Sulur block, Sarkar Samkulam block, Thudiyalur block) were collected.

The base maps were scanned and the same were inserted into AutoCad–2000. The various thematic maps like soil blocks, land use map were digitized using AutoCad–2000. This maps have been imported to ArcView. They are georeferenced, buffer was created 10 km from city to find the new site. All the maps were overlaided using Model Builder which is available in ArcView Spatial Analyst. Finally the suitable sites for dumping of solid waste disposal were found. The weighted overlay table used in the Model Builder is shown in Table 6.2. The schematic representation of analysis that was carried out using Model Builder is shown in Fig. 6.27

Advanced GIS Applications

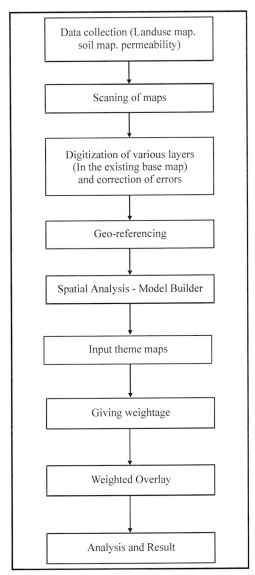

Figure 6.26: Flow chart for Methodology

6.5.3.3 Landuse or land cover

Land use or land cover may be difined as the various ways in which land may be employed or occupied. Planners compile, classify, study and analyze landuse data for many purposes, including the identification of trends, the forecasting of space and infrastructure requirements, the provision of adequate land

Table 6.2: Model Builder – Weighted overlay table

Input Theme	Label	% of influence	Weightage
Landuse	Crop Land	75%	Restricted
	Up Lands with or without scrub		5
	Mining and Industrial Land		4
	Salt effected Land		5
	Follow or Harvested Land		4
	Builtup Land		Restricted
Soil Permeability	Rapid	25%	3
	Moderately Rapid		4
	Slow		5

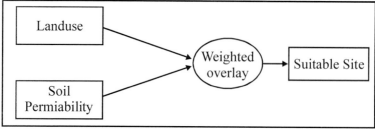

Figure 6.27: Model Builder– Schematic sketch

area for necessary types of land use, and the development or revision of comprehensive plans and land use regulations.

6.5.3.4 Permeability of Soil

Coimbatore soil profile is basically of four orders, *Entisols, Inceptisols, Alfsols,* and *Vertisols*.

Entisols

Entisols are soils without significant profile development. Usually entisols have such little profile development because they are young soils. They are often developed on bedrock, lava flows, or on alluvial (river) materials where there simply has not been time for soil formation to take place. Some entisols are found on steep slopes where run-off strips away and erodes weathering material faster than the action of soil forming processes. Permeability is rapid for Entisols.

Inceptisols

Inceptisols are soils that have "some" soil horizonal development. Basically there is an "inception" = beginning of a "B" horizon. Inceptisols are usually present where soils are relatively young or are in areas where climatic or topograpic conditions do not promote rapid soil development — semi-arid zones or on moderate slopes. Permeablity is moderately rapid in these soils.

Alfsols

Alfsols as the name implies are high in "Al" (aluminum) and "F" for Fe=iron. The enrichment in aluminum is in the form of aluminum rich clays in the "B" horizon. Mostly Alfsols have developed under "native deciduous" (hardwood) forests in cool to warm humid areas. This soil poses rapid permeability.

Vertisols

Vertisols are soils with high amounts of expanding clay. Because soils can only swell so much laterally, much of the swelling in clays such as montmorillonite is upward *i.e.* vertical and hence called as "vertisol". Swelling related to vertisols has many implications for agriculture and for environmental geology. Slow permeability is the characteristics of vertisols. The sites favourable for dumping the solid waste is shown in Fig 6.28.

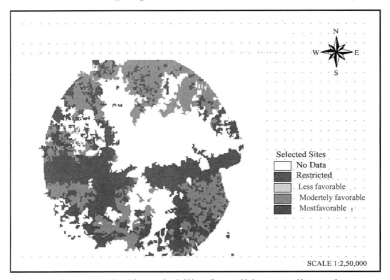

Figure 6.28: Site suitability for solid waste disposal.
(See colour version on page 185)

140   GIS: Fundamentals, Applications & Implementations

6.5.4 Groundwater Quality Assessment Using GIS For Coimbatore District

6.5.4.1 Introduction

Water is the elixir of all lives. It is the essential natural resource for sustaining our life and environment. Conservation and preservation of water resources is urgently required to be done. Over-exploitation or excessive pumping is lead to lowering of groundwater table in Coimbatore district.

As water percolates into the ground, most of the suspended solids and bacteria are removed; however, the percolating water will dissolve various minerals that come in contact with it during its passage through the soil strata. The amount and character of dissolved minerals depend on the length of the underground strata and the chemical make up of the geological formations traversed.

In this scenario, the need for maintaining the water quality standards within the permissible limits is highly demanding. The various water quality parameters and their variation over the entire Coimbatore district has been analyzed using the ArcView GIS.

Groundwater occurs from 600 feet to 900 feet in the most parts of the Coimbatore district. Extensive agriculture and industrial activities lead to this problem and in turn the water quality is also deteriorated due to the water rock interaction at higher depths. An attempt has been made to present the water quality status in Coimbatore district.

6.5.4.2 Study Area

Coimbatore district is located between latitude 10°12′N to 11°57′ N and longitude 76°39′E to 77°56′E with an areal extent of 7469 km². Groundwater is available in semi-confined to unconfined condition. Coimbatore district consists of 21 blocks and out of 21 blocks, not even a single white block (with sufficient amount of water) is available. Eleven blocks are designated as dark blocks (85% to 100% groundwater extractions) and ten blocks (65% to 85% groundwater extractions) as Grey blocks.

Coimbatore district is considered as an industrial district, but it is no way back in agriculture. Coimbatore district is famous for cotton output. Cotton occupies an important role in the economy of the district. The annual rainfall in the district ranges from 55 to 75 cm.

6.5.4.3 Materials and Methods

Survey of India toposheet in 1:2,50,000 scale for Coimbatore district was scanned as TIFF image. This image was inserted into ArcView using the TIFF reader extension available in Arcview. The boundary layer was digitized as polygon. The water sample locations were digitized as point layer. The water quality parameters like EC, TDS, pH, TH, Ca, Mg, Na, K, NO_3, SO_4, Cl, Fe, Mn, and F were added into the table for each location in the point theme.

Model Builder in the ArcView Spatial Analyst extension has been used to create a spatial model of water quality of Coimbatore district. A spatial model records the processes, such as buffering or overlaying themes, required to convert input data into an output map. Large models can be built by connecting several processes together.

Model default properties, such as the colour scheme, extent, and cell size of the output theme were set. The model was built by adding, connecting and editing processes using wizards that walks us through adding the process and defining its properties. The model was saved and was thus run to get the result.

The vector conversion process converts the point data of various water quality parameters to a grid themes. The resulting grid theme has been used as input to other process such as Reclassification and Arithmetic Overlay. All the maps were classified into two categories such as areas with desirable and undesirable water quality based on IS: 10500-1983.

The point interpolation process creates a continuous grid theme that represents a surface by interpolating the measured values in a point theme. All the layers were overlaid using Model Builder. The model has been run and cumulative score was obtained. Finally the various zones of the district were ranked with respect to their cumulative score is shown in Table 6.3.

Table 6.3: Reclassification Table – Model Builder

Class Start Value	Class End Value	Rank	Label
0	2.8	1	Very Poor
2.8	5.6	2	Poor
5.6	8.4	3	Moderate
8.4	11.2	4	Good
11.2	14.0	5	Very Good

RESULTS AND DISCUSSION

The model created using the Model Builder in ArcView has been run and the derived resultant map is shown in Fig 6.29

From the given figure the water quality status for the blocks in Coimbatore district has been interpreted and given in Table 6.4.

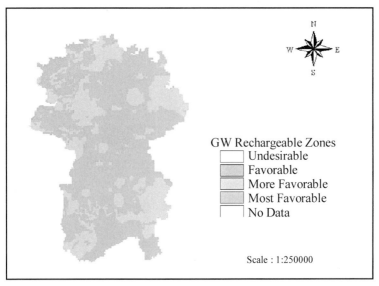

Figure 6.29: Groundwater Quality Map of Coimbatore district

Table 6.4: Block-wise Water Quality of Coimbatore District

No.	Label	Block(s)
1	Very Poor	- Nil -
2	Poor	Palladam, Tiruppur, Madukkarai, Madathukulam
3	Moderate	Sulur, Pongalur, Thondamuthur, Gudimangalam, Pollachi (N), Udumalpet, Sarkar Samakulam
4	Good	Avanashi, Sulthanpet, Pollachi (S), Karamadai, Kinathukadavu, P.N.Palayam, Anamalai, Perur, Valparai
5	Very Good	Annur

The map was interpreted for water quality such as poor, very poor, very moderate, good and very good. This map shows that Palladam, Tiruppur, Madukkarai and Madathukulam blocks posses poor groundwater quality. Annur block is found to be very good and remaining blocks posses moderate to good groundwater quality.

6.5.5 Site Suitability Analysis Using GIS For Urban Planning
6.5.5.1 Introduction

The main objective of this study is to find out the suitable site for further urban development of Coimbatore Local Planning Area using GIS.In India, due to unprecedented population growth coupled with unplanned developmental activities has led to urbanization, which lacks infrastructure facilities. Which cause serious implications on the resource base of the region.

Coimbatore popularly known as "Manchester of South India" is situated in the western part of the state of Tamilnadu. Coimbatore city is the District headquarters, where Coimbatore local planning area covers the Coimbatore Corporation,Kurichi New town, and parts of Palladam taluk, Coimbatore taluk, Avinashi taluk, Mettupalayam taluk.

Coimbatore is well known for its textile industries and has excellent potential for industrial growth. Because of its proximity to the hills of the western ghats, Coimbatore enjoys an excellent climate throughout the year. Coimabatore city was constituted as a municipality in november 1866, with a population of 24000, covering an extent of 10.88 sq.km. Coimbatore city's status has since been elevated as municipal corporation from 1.5.1981. The study area has bounded between $76^0 81'$ N to $77^0 23'$ N and $10^0 83'$ E to $11^0 25'$ E. Table 6.5 and 6.6 shows the population growth and projected population growth respectively. The expected population of Coimbatore city in 2006 is only 11.5 lakhs. but now the current population is 13 lakhs+1.5 lakhs / day floating population,which shows the immediate need of urban planning and development of Coimbatore.

To overcome the undesirable urban growth in Coimbatore LPA ,it is necessary to plan and develop the urbanization process using GIS.

Table 6.5: Population Growth

Year	Population(Lakes)
1911	0.47
1921	0.68
1931	0.95
1941	1.3
1951	1.98
1961	2.86
1971	3.56
1981	7.04
1991	8.16

Table 6.6: Projected population Growth

Year	Population(Lakes)
1998	9.71
2001	10.26
2006	11.15
2011	12.04
2016	12.9
2021	13.74

6.5.5.2 *Methodology*

The land use base map (2001) has collected. Other attributed data *viz.*, crime data has collected from District Commissioner Office and S.P. Office, school zone data has collected from C.E.O.Office, Coimbatore, Property value data has collected from District Registration office and from the website www.tnreginet.net.

From the base map, village and taluk boundaries, road maps, and land use maps were digitized in AutoCad and they were exported to GIS ArcView3.2a Software.In ArcView, Digitized maps are Georeferenced by Register and Transform tool.

In ArcView 3.2a, attribute layers are prepared *viz.*, crime data, school data, Property value data are entered into village boundary as attribute data, and created as a separate layer, for road map, the topology has been built by Data Automation Kit, the analysis has done by spatial analyst – Model Builder.

For schools and road maps, buffers zones are created with 4 buffers having a buffer width of 0.25km each. Property value, Landuse and crime data are reclassified by reclassification process. In the reclassification process, ArcView divides each theme into five parts or intervals.

In the Model Builder, in case of crime data theme high scale value has been given to the lower crime area, very low scale value will be given to the higher crime area. The number of cases will be taken as an attribute data. Similarly, in case of land rate data theme, high scale value has been given to the land, which is estimated as lower square feet rate, very low scale value will be given to the land which is estimated as higher square feet rate. Similar to that, high scale value has been given to the waste land or agricultural land and low scale value will be given to the other industrial, commercial, and public, semi public areas. Crime map (Fig 6.30), property value map (Fig 6.31), landuse map (Fig 6.32), road map (Fig 6.33), school zone map (Fig 6.34) have been created. Finally arithmetical overlay has been done by spatial analyst.

Suitability

Score = School (M) + Crime (M) + Road (M) + Landuse (M)+ Property Value (M)

M – Multiplier

Crime theme – 2 School theme – 1.5 Road buffer theme – 2
Property value theme – 3   Landuse theme – 1.5

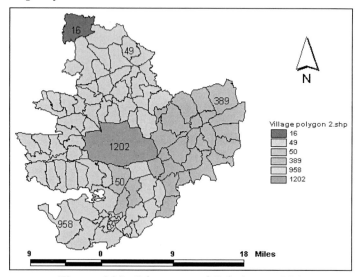

Figure 6.30: Crime map of Coimbatore LPA

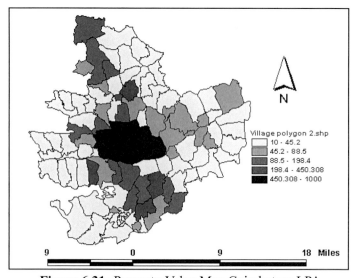

Figure 6.31: Property Value Map Coimbatore LPA

Advanced GIS Applications 147

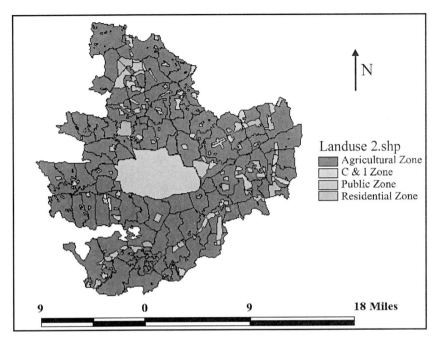

Figure 6.32: Landuse Map of Coimbatore LPA

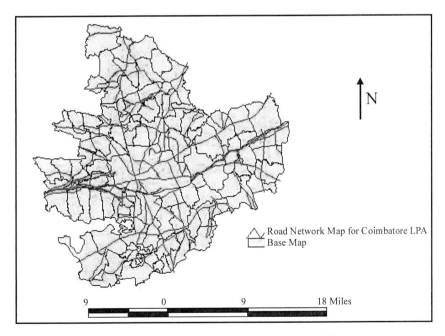

Figure 6.33: Road Map of Coimbatore LPA

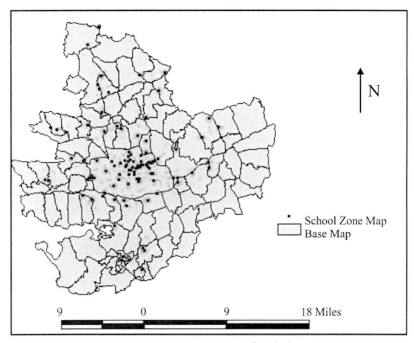

Figure 6.34: School Zone Map of Coimbatore LPA

Suitability Score

Suitability score = School (1.5) + Crime (2) + Road (2) + Landuse (1.5) + Property Value (3)

The range of suitability score has equally divided into five parts, these are 35.2 – 44 as very suitable, 26.4 – 35.2 as suitable, 17.6 – 26.4 as moderately suitable, 8.8 – 17.6 as less suitable, 0.0 - 8.8 as unsuitable. The existing urban area has been given to a high scale value in order to find the suitable site for further urban development of Coimbatore. Mode (Fig 6.35) has been run and the result is shown in Fig 6.36.

6.5.5.3 Result

In this study, it is found that the best suitable area is lying towards the middle south of the Coimbatore Corporation which is nothing but a Kurichi New Town and Karunjami Kaundan Palayam, Sundakkamuthur and Maddukkarai in Coimbatore taluk.

Advanced GIS Applications

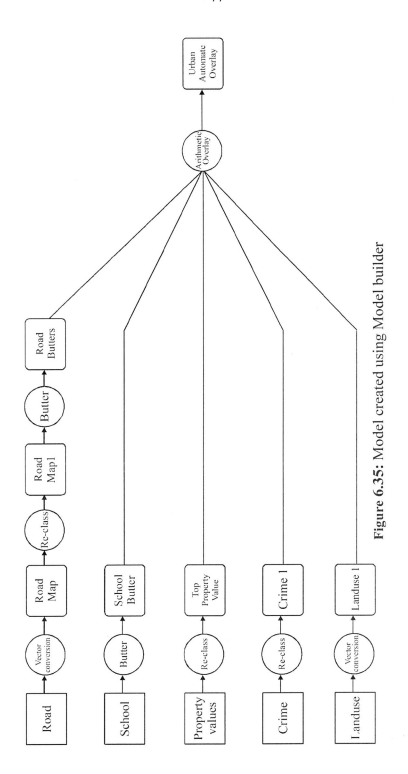

Figure 6.35: Model created using Model builder

Some areas lying towards north of the city which covers Kuppipalayam, Kattampatti, Kunnathur in Avinashi taluk and Bilichi (Around Mettupalayam Road), Narasimma naicken palayam, Idhikarai, Agraharasamakulam, Vellanaipatti and Kallipalayam in Coimbatore taluk.

Some areas lying towards west of the city, which covers Veerakeralam, Thennamanallur, Kalicka naicken palayam and Devarayapuram in Coimbatore taluk.

Some areas lying towards East of the city are Nilambur, Sulur, Rasipalayam, Maileripalayam and Kannampalayam in Coimbatore taluk.

Some areas lying towards North East of the city which covers Karumathampatti, Kaniyur, Arasur, regions in Palladam taluk.

Some areas lying towards South East of the city which covers Appanaickenpatti, Pappampatti in Palladam taluk, and Pachapalayam, Orathukuppai, Ottakkalmandapam, in Coimbatore taluk.

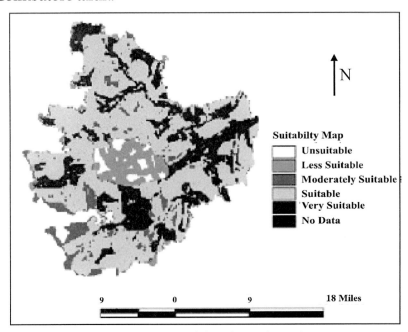

Figure 6.36: Site suitability for Urban Planning

Some areas lying towards North West of the city, which covers Maruthur region in Mettupalayam taluk and Naickenpalayam in Coimbatore taluk.

Most suitable area is lying in most of the areas of Coimbatore LPA due to the presence of agricultural zones. The Coimbatore Corporation is covered under the less suitable and unsuitable areas due to the high scale value given to the land use theme in model builder. Moderately suitable area is lying in some small part of Coimbatore, Avinashi, and Palladam taluks.

6.5.6 Soil Erosion Modeling For Coimbatore District
6.5.6.1 Introduction

Soil erosion is a silent process of the nature. The effect of soil erosion is a progressive one. Agriculture and construction industries are directly related to the available soil mass. Deforestation and all other anthropogenic activities cause the total soil loss. The on-site and off-site effects of soil erosion attract us towards the prevention of soil erosion. Soil erodibility of different parts in the Coimbatore district have been found using Geographic Information System.

Soil erosion is one form of soil degradation along with soil compaction, low organic matter, loss of soil structure, poor internal drainage, salinisation, and soil acidity problems. These other forms of soil degradation are serious in themselves and usually contribute to accelerated soil erosion. Soil erosion is a naturally occurring process on all types of lands. The agents of soil erosion are water and wind. In a recent overview of global erosion and sedimentation it is stated that more than 50 % of the world's pasturelands and about 80 % of agricultural lands suffer from significant erosion.

According to information published by Ministry of Agriculture (Govt.of India) in 1980,in India as many as 175 Mha (constituting 53 % of India's geographical area) are subject to environmental degradation. About 150 Mha has been caught in the vicious circle of erosion by water and wind. One estimate puts the loss of topsoil by the water action at 12000-m tonnes

152   *GIS: Fundamentals, Applications & Implementations*

every year. Soil erosion may be a slow process that continues relatively unnoticed, or it may occur at an alarming rate causing serious loss of topsoil. The loss of soil from farmland may be reflected in reduced crop production potential, lower surface water quality and damaged drainage networks.

6.5.6.2 Factors Affecting Soil Erosion

Rainfall Intensity and Runoff

Both rainfall and runoff factors must be considered in assessing a water erosion problem. The impact of raindrops on the soil surface can break down soil aggregates and disperse the aggregate material. Lighter aggregate materials such as very fine sand, silt, clay and organic matter can be easily removed by the raindrop splash and run-off water; greater raindrop energy or run-off amounts might be required to move the larger sand and gravel particles.

Soil movement by rainfall (raindrop splash) is usually greatest and is most noticeable during short-duration, high-intensity thunderstorms. Although the erosion caused by long-lasting and less-intensive storms is not as spectacular or noticeable as that produced during thunderstorms, the amount of soil loss can be significant, especially when compounded over time. Run-off can occur whenever there is excess water on a slope that cannot be absorbed into the soil or trapped on the surface. The amount of run-off can be increased if infiltration is reduced due to soil compaction, crusting or freezing. Run-off from the agricultural land may be greatest during spring months when the soils are usually saturated, and vegetative cover is minimal.

Soil Erodibility

Soil erodibility is an estimate of the ability of soils to resist erosion, based on the physical characteristics of each soil. Sand, sandy loam and loam-textured soils subject to be less erodible than silt, very fine sand, and certain clay textured soils.

Past erosion has an effect on soil erodibility for a number of reasons. Many exposed sub-surface soils on eroded sites tend to be more erodible than the original soils were, because of their poorer structure and lower organic matter. The lower nutrient levels often associated with sub-soil contribute to lower crop yields and generally poorer crop cover, which in turn provides less crop protection for the soil.

Slope Gradient and Length

Naturally, the steeper the slope of a field, the greater the amount of soil loss from erosion by water. Soil erosion by water also increases as the slope length increases due to the greater accumulation of run-off.

Vegetation

The potential of the soil to erode increases if, the soil has no or very little vegetative cover of plants and/or crop residues. Plant and residue cover protects the soil from raindrop impact and splash, which tends to slow down the movement of surface run-off and allows excess surface water to infiltrate.

The erosion-reducing effectiveness of plants and/or residue covers depends up on the type, extent and quantity of cover. Vegetation and residue combinations that completely cover the soil, which intercepts all falling raindrops close to the surface and is the most efficient method in controlling soil. Partially incorporated residues and residual roots are also important as these provide channels that allow surface water to move into the soil.

6.5.6.3 Methodology

The base map of the Coimbatore district was scanned and the same was imported into AutoCad-2000. The various thematic maps like landuse, soil and slope were digitized using AutoCad-2000. These maps were exported into ArcView GIS Rainfall map has been prepared based on rainfall data. All the thematic maps were given ranking and overlaid using Model Builder available in ArcView Spatial Analyst. Finally the erodibility of different parts of the district was arrived.

154   *GIS: Fundamentals, Applications & Implementations*

The reclassified values for the layers are given in Table 6.7. All these layers are overlaid using Arithmetic Overlay Operator. The resultant map is reclassified as the values given in Table 6.8.

Table 6.7: Reclassified values for layers.

Input Theme	Operator	Multiplier	Label	Value
Rainfall	+	1	4.75 - 13.72	1
			13.72 - 22.69	3
			22.69 - 31.66	5
			31.66 - 40.63	7
			40.63 - 49.6	9
			No Data	0
Slope	+	4	Plain	1
			Very Low	3
			Low	5
			Moderate	7
			High	8
			Very High	9
			No Data	0
LandUse	+	3	Arable Land Unirrigated	9
			Arable Land Irrigated	7
			Forest	1
			Mines	2
			Urban Settlements	5
			Water Bodies	0
			No Data	0
Soil	+	2	Rock Outcrops	1
			Typic Humitropepts	2
			Typic Ustropepts	2
			Entic Chromusterts	3
			Lithic Ustrothents	4
			Typic Haplustalfs	4
			Typic Rhodustalfs	4
			No Data	0

Table 6.8: Model Builder- Reclassification

Class Start Value	Class End Value	Rank	Label
0	16	1	Very Much Safe
16	32	2	More Safe
32	48	3	Moderate
48	64	4	Dangerous
64	115	5	Highly Dangerous

6.5.6.4 Conclusion

Soil erosion can be controlled effectively if it is predicted accurately under alternate management strategies and practices. The results of the study shows that most of the parts of the Coimbatore district are quite safe against soil erosion. The final results are shown in Fig. 6.37. This study concentrates only on the soil erosion by water. This work can be further furnished by including the other themes like soil roughness, climate *etc.*, to account for the soil erosion by wind. The concept of this study can be adopted for different parts of the country especially in hilly areas to the high risk zones and proper measures can be taken to prevent soil erosion.

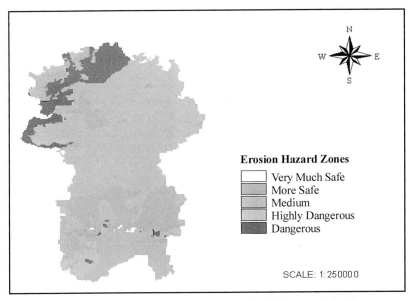

Figure 6.37: Erosion Hazard Map of Coimbatore District.

156   *GIS: Fundamentals, Applications & Implementations*

6.5.7 Groundwater Level Variation Level Analysis in Coimbatore District

6.5.7.1 Introduction

Groundwater is precious in drought regions. Annual rainfall of Coimbatore is relatively low as compared with other parts of Tamil Nadu. Also due to intensive agricultural and industrial activities, groundwater level has been lowered to the depth of 30m at few places. Hence it is important to monitor the groundwater level of the district.

6.5.7.2 Methodology

The groundwater table levels of Coimbatore district during 1991-2003 have been collected. The base map of Coimbatore district was scanned and the same was imported into AutoCad-2000.The boundary of the Coimbatore district and the locations for which data is available are digitized using AutoCad-2000. These maps were exported into ArcView GIS 3.2a .The collected groundwater table levels were entered for the locations and the values were interpolated for the year 1991 and 2003. Table 6.9 shows the exploitation level of groundwater in Coimbatore district. Finally, groundwater level in 1991 was compared with 2003 and the analysis was carried out in Model Builder. The difference in variation of groundwater table between May 1991 and May 2003 is shown in Fig. 6.38.

Table 6.9: Block-wise Groundwater Table Level variation in Coimbatore District (1991-2003)

No.	Label	Blocks
1	Good	Tiruppur, Avanashi, Madukkarai, Pongalur, Pollachi(N), Madathukulam
2	Exploited	Palladam, Sulur, Annur, Sulthanpet, Thondamuthur, Gudimangalam, Pollachi(S), Karamadai, Kinathukadavu, Periyanayakkanpalayam, Sarkar Samakulam
3	Over Exploited	Udumelpet, Anamalai

6.5.7.3 Conclusion

The results of the study shows that groundwater in most parts of the district were exploited fully. This study shows that groundwater table level varies abruptly in Coimbatore district. Water demand in the district keeps on increasing with the tremendous increase in population. In this alarming situation it is important to monitor the variation of groundwater table level in Coimbatore district otherwise there will be acute shortage for groundwater in the forthcoming years.

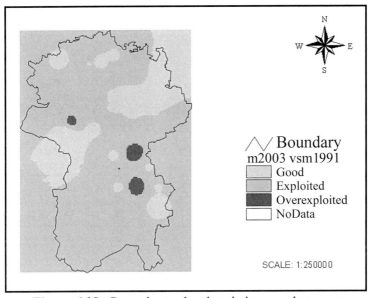

Figure 6.38: Groundwater level variation map between 1991 and 2003 for Coimbatore District.

6.5.8 Groundwater Quality Assessment Using GIS In Coimbatore District

6.5.8.1 Introduction

The usage of groundwater for agricultural, domestic and industrial purposes is determined based on the groundwater quality. The standards for these purposes are also different. Hence a study has been made to asses the overall groundwater quality of Coimbatore using GIS. The need for maintaining the water quality standards within the permissible limits is highly demanding.

158   *GIS: Fundamentals, Applications & Implementations*

6.5.8.2 *Methodology*

The base map of the Coimbatore district was scanned and the same was inseted into AutoCad –2000. The boundary of the district and the various locations for which data is available are digitized and then these layers were imported into ArcView GIS 3.2a. The water quality data for various locations have been entered and the data were interpolated for various water quality parameters. Table 6.10 shows the permissible and excess limits proposed by IS: 10500-1983, Indian Council for Medical Research and World Health Organization. Using the Indian Water Quality standards (IS:10500-1983), the overall Coimbatore district was subdivided into desirable and undesirable zones for various water quality parameters.

Table 6.10: Standards for drinking water

Parameter	INDIAN STANDARDS 10500-1983		ICMR		WHO	
Physical	P	E	P	E	P	E
Turbidity	10	25	5	25	5	25
Chemical	P	E	P	E	P	E
pH	6.5-8.5	6.5-9.2	7-8.5	6.5-9.2	7-8.5	6.5-9.2
Total solids	-	-	-	-	500	1500
Total Hardness	300	600	300	600	-	-
Calcium	75	200	75	200	75	200
Magnesium	30	100	50	150	50	150
Iron	03	1.0	.3	1.0	.3	1.0
Manganese	0.1	0.5	0.1	0.5	0.1	0.5
Chlorides	250	1000	250	1000	200	600
Sulphates	150	400	200	400	200	400
Nitrate	45	-	20	50	-	50-100
Na + K	-	-	-	-	200	-
Fluoride	0.6-1.2	-	1.0	2.0	0.5	1.0-1.5

P = PERMISSIBLE LIMIT   E = EXCESSIVE LIMIT
NOTE: All units are in mg/l except pH

Finally all the layers were overlaid using Model Builder. The results are shown in the figure 6.39 Finally the various zones of the district were ranked with respect to their cumulative score as shown in Table 6.11.

Table 6.11: Reclassification Table – Model Builder

Class Start Value	Class End Value	Rank	Label
0	2.8	1	Very Poor
2.8	5.6	2	Poor
5.6	8.4	3	Moderate
8.4	11.2	4	Good
11.2	14	5	Very Good

6.5.8.3 Conclusion

The results of the study shows that many parts of Coimbatore district failed to satisfy atleast one water quality parameter standard. The results also shows that each part of the district satisfies atleast 5 water quality parameter standards out of 15 considered. The block-wise water quality of Coimbatore district is shown in Table 6.12. Annur is the only block labeled as very good. The overall water quality of Coimbatore district is

Table 6.12: Block-wise water quality of Coimbatore district

No.	Label	Block(s)
1	Very Poor	- Nil -
2	Poor	Palladam, Tiruppur, Madukkarai, Madathukulam
3	Moderate	Sulur, Pongalur, Thondamuthur, Gudimangalam, Pollachi(N), Udumalpet, Sarkar Samakulam
4	Good	Avanashi, Sulthanpet, Pollachi(S), Karamadai, Kinathukadavu, P.N.Palayam, Anaimalai
5	Very Good	Annur

not so bad. This study will be helpful in carrying out water quality analysis using GIS for the other parts of the country. Heavy industrialization and urbanization may be the basic causes of water pollution. Strict rules may help to prevent further water pollution caused by industries. Large scale of awareness must be provided to the society about the need of water quality maintenance.

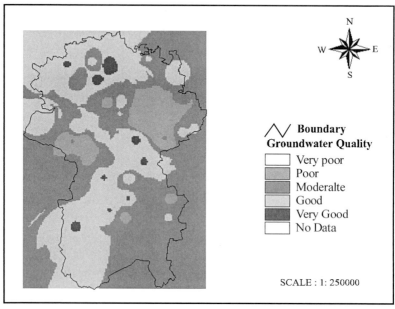

Figure 6.39: Groundwater Quality Map of Coimbatore district

6.5.9 Rejuvenation of Noyyal River Using GIS

6.5.9.1 Study Area

The river Noyyal originates in the hills of Western Ghats towards the south-west of Coimbatore. The river course is almost 172km long to join the river Cauvery, at Kodumudi, Karur district (latitudes: 10° 54' 00" and 11° 19'03" N and longitudes 76° 39'30" and 77° 05'25" E). During its course Noyyal flows through Coimbatore, Erode and Karur districts. The Noyyal basin covering 3510 sq. km, out of this 49.9% of the area is under cultivation and 178 km^2 (5.1% of the total area under forest and wasteland growing teak and eucalyptus). The rest 45%(1580 sq. km) is barren, uncultivated lands, rocky strata, permanent

Advanced GIS Applications 161

pastures and fallowlands. Fig 6.40 shows the Noyyal river basin map.

Altitude of the first order streams, the tributaries of Noyyal originating in the Western Ghats catchments, ranges upto 1200m. In the central and eastern parts and towards the confluence of the river with the river Cauvery the terrain is plain and altitude is around 200 m. Based on the topography, the river can be segmented into three parts, (i) valleys and mountains, the slopes in Western Ghats, a well-grained region, (ii) fairly plain flat, lower catchment area and, (iii) East Noyyal river basin with insignificant slope.

The river was perennial with good flow till early seventies. In recent years, the scene has been changed drastically and the river has become practically seasonal. River Noyyal receives copious water during northeast monsoon from September to November. The rest of the year it remains more or less dry.

The river Noyyal, in recent decades has transformed into a seasonal river. The average rainfall in the basin is about 700 mm. This river is the only source for 31 tanks and many minor canals. All the supply canals and tanks are having a command area of 6550 hectares. Water table has been lowered even upto 300 m in many places in Noyyal river basin due to the over exploitation of groundwater by borewells for agricultural and industrial purposes.

6.5.9.2 Data for the Study

The Indian Remote Sensing Satellite (1RS-1D) digital data of LISS-III sensor, which provide a resolution of 23.4 meters in multispectral mode was utilized in the study. The details of satellite data used in the study are given below:

S.No.	Path/Row for IRS- 1D, LISS –III Digital Satellite Data	Date of Pass
1.	100-66	22, December 2002

Survey of India (SOI) topographical sheets of Coimbatore district on 1:50,000 scale were utilized for registration of satellite data, selection of ground control points and locating training

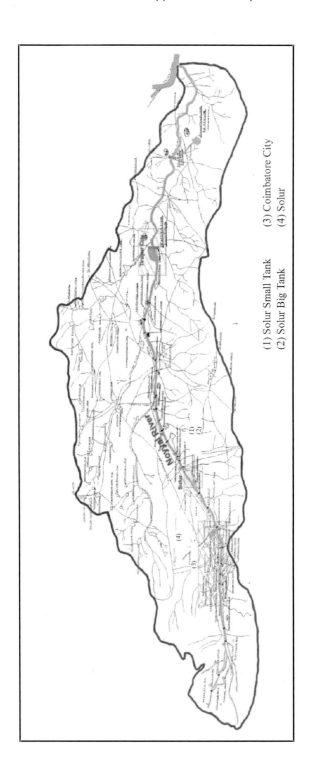

Figure 6.40: Noyyal River Basin Map

Advanced GIS Applications

sets as well as to identify and authenticate the various locations on the satellite image.

Noyyal river basin map of 1:125000 scale was also utilized to demarcate the study area on the satellite images.

The characteristics of the data used in the study is given as follows:

Liss-III

Resolution	:	23.4 m
Swath	:	127 km (bands 2, 3, 4)
	:	134 km (band 5 –MIR)
Repetitivity	:	24 days
Spectral Bands	:	0.52 - 059 microns (B2)
	:	0.62 - 0.68 microns (B3)
	:	0.77 - 0.86 microns (B4)
	:	1.55 - 1.7 microns (B5)

6.5.9.3 Tributaries of river Noyyal

The Noyyal River has seven major tributaries (Table 6.13). Each of these streams is formed from a number of first order and second order streams that drain own micro watersheds

Table 6.13: The major tributaries of Noyyal river basin.

Name of the Stream	Direct Irrigation (in acres)
Koduvaipudi Odai	191.10
Mullurambu Odai	27.94
Mudanthurai Odai	43.63
Irruttupallam Odai	8.76
Sundaram Odai	26.54
Pachaan Vaikal	173.53
Kanchi Manathi	1554.77

All these tributaries and their lower order streams start from the foot of the Western Ghats. On river Noyyal along its course a number of anaicuts are present. The first Anaicut is chithiraichavadi Anaicut, which is currently almost 95% silted, that has reached the crest level. All the streams get flooded during rainy seasons, damaging the banks. Since the catchments

164   GIS: Fundamentals, Applications & Implementations

area at the upper reaches is steep with high slope, water flows with great velocity causing serious erosion.

6.5.9.4 Anaicuts of Noyyal basin

For the purpose of irrigation 23 major anaicuts (Table 6.14) were built in river Noyyal during various periods in the past. Most of the dams currently maintained and are in very dilapidated conditions.

Table 6.14: Anaicuts and Irrigated areas in the Noyyal river basin.

S.No.	Name	Irrigated Area in acres
1	Chithiraichavadi anaicuts	3858
2	Kuniamuthur anaicuts	2093
3	Coimbatore anaicuts	2559
4	Kurichi anaicuts	509
5	Vellalore anaicuts	617
6	Singanallur anaicuts	1318
7	Ottarpalayam anaicuts	822
8	Irrugur anaicuts	409
9	Sulur anaicuts	705
10	Rasipalayam anaicuts	320
11	Madappur anaicuts	101
12	Samalapuram anaicuts	155
13	Karumathampatti anaicuts	68
14	Pallapalayam anaicuts	118
15	Semmandampalayam anaicuts	91
16	Akkiraharapudur anaicuts	Damaged
17	Mangalam anaicuts	205
18	Andipalayam anaicuts	Damaged
19	Tirupur anaicuts	50
20	Mannarai anaicuts	59
21	Mudalipalayam anaicuts	170
22	Anaipalayam anaicuts	132
23	Kathankanni anaicuts	225

Apart from these check dams, Neeli and Pudukkadu dams are constructed over Kanchimanathi, the first major tributary, to irrigate 1555 acres. Agrahar Puduppalayam and Thamari Kulam dams are constructed over Nallar River and irrigate nearly 554.92 acres. Sambamadai dam and 10 small ponds *i.e.,*

Thoravalur, Athur, Mandrakkanai, Kunnathur, Vellarivelli, Kanchi Koundampalayam, Puthur, Punjaipalatholuvu, Thalavai-palayam and Thuvakkamuthur in Avarakkari Pallam irrigates nearly 696.74 acres.

For direct irrigation, 30 ponds with 14 canals were constructed which irrigates 9331 acres. Apart from these larger ponds there are 13 small ponds getting water for irrigating 1214.45 acres. Apart from the major check dams, there are 55 small check dams in the entire river course of Noyyal from its origin. Using these water harvesting structures 138 million cubic water is stored which helps indirectly for well irrigation.

The Noyyal river basin is having 23 anaicuts and 30 tanks are constructed to facilitate the direct irrigation purpose.

Soil classification map of the basin shows six different types of soil and they are as follows:

— *Red calcareous soil*

— *Black soil*

— *Red non-calcareous soil*

— *Alluvial and Colluvial soil*

— *Brown soil*

— *Forest soil*

The major encroachments in the Noyyal river basin mostly are on the tanks and on both side of the Anaicut supply channel area. The encroachment is very high in Coimbatore and Tirupur town area. The encroachments are present almost in all tanks and channel up to Mannarai area. After Mannarai there are no encroachments. In all the encroachments mentioned bellow, the huts are provided with drinking water connection and all other infrastructure facilities.

The study aims to ascertain the existing condition of the Noyyal river basin by comparing the past and the present conditions through field studies satellite images and also to quantify the encroachment in lakes.

166   *GIS: Fundamentals, Applications & Implementations*

1. To generate a database having the information of ground water level and rainfall in the Noyyal river basin and to find the places with deeper groundwater level so that appropriate steps may be taken to augment water level.
2. To analyze the water quality level at different selected locations in Noyyal river basin using geographic information system.
3. To analyze and to propose suitable measures to revive the Noyyal river.

6.5.9.5 Methodology

1. Noyyal river basin map was digitized in AutoCad and saved it in .dxf format. This .dxf file was been then export into ArcView.
2. Projection for the digitized map was set as UTM WGS-84 NORTH and Zone 43.
3. Georeferencing was done using control points taken from Survey of India topographical sheets in 1: 50,000 scale.

ERDAS IMAGINE 8.4 image processing software has been used in this study.

LOCATING GROUND CONTROL POINTS

This process employs identification of geographic features on the image called Ground Control Points (GCPs), whose positions are known as intersection of streams, highways, airport, runways *etc*. Longitude and latitude of GCPs can be determined by using accurate base maps or from GPS points. Accurate GCPs are essential to accurate rectification. GCPs should be reliably matched between source and reference (*e.g.*, road intersection) and they were selected such that they are widely dispersed throughout the source image.

IMAGE ENHANCEMENT TECHNIQUES

Various image enhancement techniques like density slicing, contrast stretching, liner contrast stretching, histogram equalization, Principal Component Analysis have been performed on the image to acquire more information.

The image has been re-projected to UTM WGS- 84 North and zone as 43 East to use the image along with the base map of Noyyal river basin.

PERFORMING GEOREFERENCING

The reprojected image was georeferenced using the already georeferenced vector layer of Noyyal basin map. The locations of the GCPs are shown in the image 6.41.

SUBSETING THE IMAGE

After georeferencing, the Noyyal basin map in vector layer was used to sub set the image to obtain the image of Noyyal basin.

CLASSIFICATION

Both supervised and unsupervised classifications were performed. As supervised classification is more useful to extract more information, it has been used for estimating the river length, landuse classifications. Fig 6.42 show part of the image after supervised classification.

FILTERING

3x3 Low Pass Filter, 3x3 High Pass Filter, 3x3 Vertical Edge Detector, 3x3 Horizontal Edge Detector, Compass filters have been used to remove the noises and obtain the more information.

6.5.9.6 Results From Erdas Imagine 8.4 Analysis

Using the IRS-1D, LISS III Satellite image the Noyyal river basin was studied and the details regarding the tanks in the Noyyal river basin, links between the tank and Noyyal river course was analyzed. The area of tanks and their perimeters were found using ERDAS Imagine 8.4 software. A comparative study has been made about the present condition of the tanks and links in the river course. Also the tanks which have been encroached by the people and the extent of encroachment were found from

168 GIS: Fundamentals, Applications & Implementations

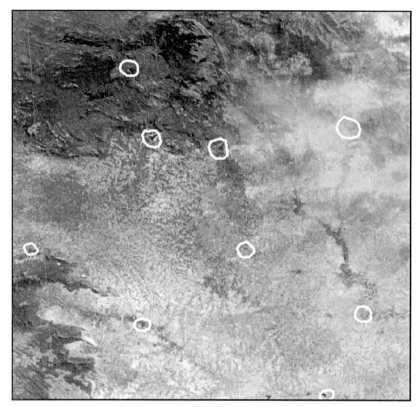

Figure 6.41: Locations of Ground Control Points.
(See colour version on page 186)

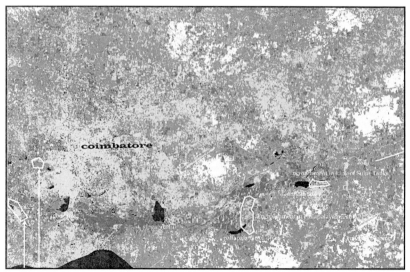

Figure 6.42: Map after supervised classification.
(See colour version on page 186)

Advanced GIS Applications   169

comparative study made on satellite image and existing past data.

From these results it is inferred that the tanks in and around Coimbatore (Fig. 6.43) and Tirupur (Fig 6.44) regions are encroached to a greater extent.

From the ERDAS Imagine 8.4 analysis the length of links between the tanks and the Noyyal river course were found. Table 6.15 shows the links of tanks, which are encroached, to greater extent by the developing city.

Table 6.15: Encroached Tanks and the encroached length of links taken from IRS-1D Satellite image.

Tank Name	Length of Link
Pallapalayam tank	1.6704
Sulur small tank	3.9633
Sulur big tank	1.4672
Mudalipalayam tank	1.0559
Annaipalayam tank	0.9437
Katanganni tank upstream link	2.2965
Katanganni tank downstream link	0.7660
Punjai palatholuvu tank	3.7759
Singanallur tank	0.8691
Coimbatore tank	1.3187
Irugur tank	1.2142

WATER SPREAD AREA AND PERIMETER OF TANKS IN NOYYAL RIVER COURSE AS TAKEN FROM IRS-1D LISS III SATELLITE IMAGE

Using ERDAS Imagine 8.4 software the water-spread area and perimeter of the tanks in Noyyal river basin were found (Table 6.16). From the study it is found that Sirear Periyapalayam tank is having a large water spread area of 0.957 sq.km and few tanks namely Selvasinthamani tank, Papankulam tank *etc.*, were found dry.

170   GIS: *Fundamentals, Applications & Implementations*

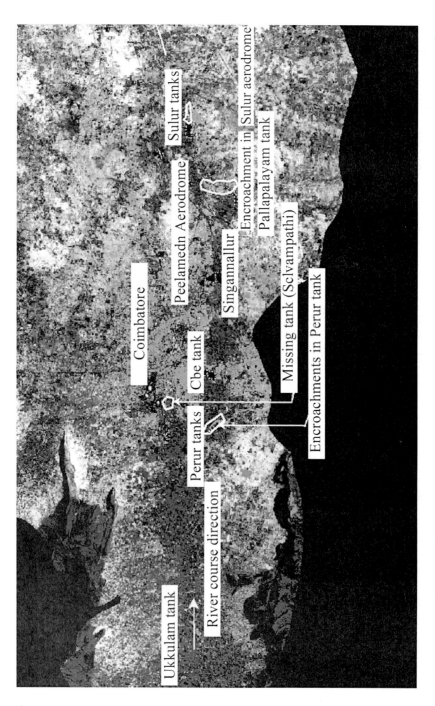

Figure 6.43: Image showing Coimbatore region. (See colour version on page 187)

Advanced GIS Applications 171

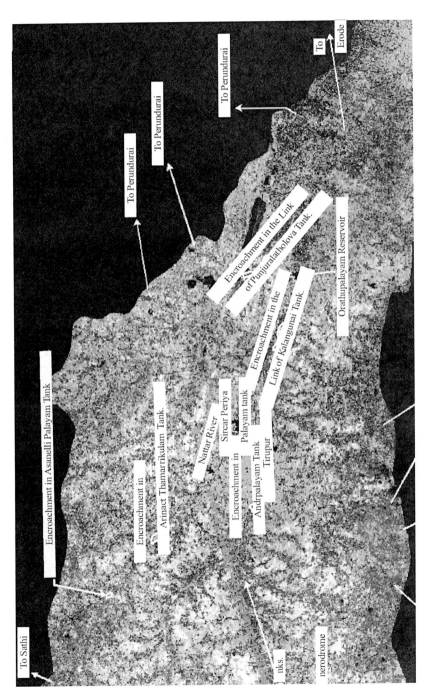

Figure 6.44: Image showing Tirupur region. (See colour version on page 188)

172   *GIS: Fundamentals, Applications & Implementations*

Table 6.16: Water spread area and perimeter of the tanks

Name of the Tanks	Water Spread Aera in sq.km	Perimeter of Tank in km
Singanallur	0.7424	8.901
Pallapalayam	0.2187	3.673
Irrugur	0.0832	1.593
Sulur big tank	0.3495	3.274
Sulur small tank	0.0452	1.498
Nilambur	0.0894	2.205
Sirear periyapalayam tank	0.9570	7.114
Anaipalayam tank	0.0344	1.726
Mudalipalayam	0.0543	1.039
Katangani tank	0.2198	2.474
Orathupalayam	1.4122	12.034
Narasampathi tank	0.1309	2.802
Coimbatore tank	0.3937	4.432
Selvasinthamani tank	0.0335	0.932
Kumarasamy tank	0.0276	0.694
Krishnampathi tank	0.0229	0.928
Pudhukulam tank	0.0237	0.801
Ukkulam tank	0.0106	0.525
Perur big tank	0.0259	1.112
Mannari tank	0.0096	0.595
Palavaipalayam tank	0.0426	1.211
Punjai palatholuvu tank	0.0605	2.175
Athur tank	0.0183	0.655
Muriyandampalayam tank	0.2408	2.881
Sarkar smakulam tank	0.0762	1.590
Vallankulam tank	0.0249	0.855
Ammankulam tank	0.0208	1.135

Advanced GIS Applications

6.5.9.7 Comparative Study on Encroachments

The results obtained from ERDAS Image 8.4 analysis and the past were compared and the area of encroachment in the tanks were found (Table 6.17). This study shows that the tanks in the Coimbatore and Tirupur regions are encroached to larger extent.

Table 6.17: Encroached area of the tanks

Name of the Tank	Encroached Area in sq.km
Singanallur tank	0.4106
Irrugur tank	0.1758
Sulur big tank	0.0189
Sulur small tank	0.1168
Neelambur tank	0.2016
Narasampathi tank	0.3710
Coimbatore big tank	0.9013
Selvasinthamani tank	0.1166
Kumarasamy tank	0.3524
Krishnampathi tank	0.6891
Pudhukulam tank	0.0714
Mannari tank	0.1524
Vallankulam tank	0.6231

From the analysis it is found that Coimbatore big tank, Krishnampathi tank and Vallankulam tank in Coimbatore region are encroached to greater extent.

6.5.9.8 Results From ArcView 3.2a Analysis

Using the ArcView 3.2a software the groundwater quality in Noyyal river basin has been analyzed based on the parameters, pH, total dissolved solids, electrical conductivity, chloride, sulphate, pottasium, calcium, hardness, magnesium, sodium. The areas having the very poor quality ground water and having good quality ground water were found.

From the Model Builder weighted overlay analysis, the areas in the Noyyal river basin are divided into four zones such as hazardous, not acceptable, permissible, good based on their groundwater quality (Fig. 6.45).

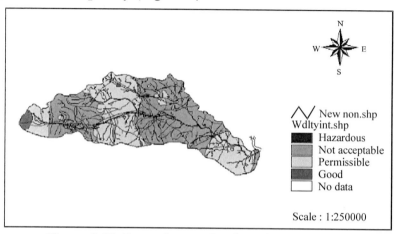

Figure 6.45: Groundwater Quality of Noyyal Basin.

The areas at the origin of Noyyal river *i.e.*, Vellingiri hills are having good quality of groundwater and with the development of cities in river course the pollution also increased. This is clearly seen from the results, the area in and around Coimbatore and Tirupur are having poor groundwater quality.

The present condition of Noyyal river basin was studied using IRS-1D satellite image and the status of the tanks in the Noyyal river course was analyzed. From the study it is inferred that the tanks in and around Coimbatore and Tirupur cities are encroached to larger extent. The links between the tanks and river course in the developing areas of Coimbatore and Tirupur cities are obstructed due to the encroachment by the people, growing of crops and development of forest.

In general, the eastern portion beyond Tirupur and Coimbatore cities are polluted much, which is due to the discharge of industrial effluents into the river in the city areas. The central portion, extreme east and extreme western regions are having good water quality due to the self-purification of river Noyyal.

Internet GIS 7

7.1 INTRODUCTION

Web technology revolutionised information technology and GIS of no exception. Analysis of maps available in internet through remote personal computer is called as web GIS. The term web is GIS synonymous with the term internet GIS. Most of the software vendors in the field of GIS have developed their GIS compatible to internet. The advantage of internet GIS is no need to own a GIS software in the personal computer. Maps available in a web site can be analysed for GIS functions as one could do with GIS in standalone personal computer. Applications of web GIS is not limited. Most of the governmental agencies, business people and others put their data in their server and the data could be analysed spatially for the users requirement. Internet mapping is simple, cost effective and reach vast audience.

Internet GIS is useful in field like civil engineering, government organisations, forestry, environment, goescience, utility planning *etc*.

7.2 HARDWARE AND SOFTWARE

Personal computer (preferably PIV) with Internet connection any one of the following softwares.

Internet GIS software are Autodesk Map Guide from Autodesk, ArcIMS from ESRI, MapExtreme Java from Mapinfo, Geomedia web map from Intergraph.

Principles of Internet GIS

The computer connected with any one of the web browser and getting into the web site which consists of web GIS. The web GIS page consists of maps with information. Many Graphical User Interface tools like pan, zoom, distance measurement tool *etc* are available. When the user zoom a map, the map is zoomed into the zoom level selected. Also many queries can be performed. An example of query is to find 5 star hotels within 5 km radius of a city. The software will perform the query and hotels within 5km will be displayed. Likewise many applications shall be performed for various applications. Both spatial and attribute data can be queried. Buffers can be created around point, line and polygon features. The result can be added with test, scale and legend and can be printed.

Client sends a request through internet from his computer by using HTML/DHTML or by Java Applet. The Web server receives the request and passes in to Coldfusion/ASP servers, which in turn hand over the request to a connector and the connector opens a path for the respective Internet GIS application server to respond. This request is sent to the application server. The application server sends the request to an available spatial server within a virtual server group. The spatial server generates the response and returns to the client. Fig 7.1 show the work process of active controls.

The problem with internet GIS is downloading which causes delay.

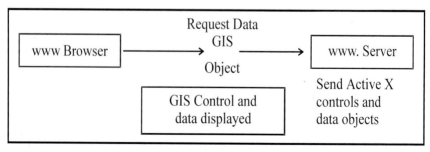

Figure 7.1: Work process of Active X controls

7.3 INTERNET GIS ARCHITECTURE

Web GIS are useful in the field of E-Commerce, Enterprise Resource Planning (ERP), data warehousing, customer care support, location based services and field data integration. The advantage of using web GIS is that no experience is required to do GIS analysis. Users can enter value or choose frame values from a pop up munu or list box or comb box and the result is obtained. Most of the internet GIS softwares support various data sources like shape file, BMP, JPEG, GIF, BIL, LAN, TIFF and MrSID formats. Web GIS are useful to publish maps in internet.

7.4 INTERNET GIS APPLICATIONS

Real time weather analysis is possible. Sensors are immersed in water body and the water quality parameters are analysed and the analysed data is sent to the map which is available in GIS server. Also weather sensors may be attached to a pole and the wind speed, temperature, pressure can be recorded and the data is sent to the GIS server. Such data shall be used by the people who are in remote place. These data very much required for utility, maintenance and planning.

7.5 FUTURE OF INTERNET GIS

Web GIS is being used now by many organisations. In future most of the government, non-government and private people will use web GIS to publish their data enable it to reach to the public. If a city map is posted in web GIS with details like important locations, like bus stops, temples, churches, mosques, tourist important places, airports, railway stations *etc.*, user shall query the map according to his interest.

Following are some of the questions that may be queried in data available, if a project is really implemented.

1. What is the shortest path between airport to city bus stand?
2. Where are 5 star hotels within 3 km radius of Railway junction?

3. What is the soil type in the locations of my interest?
4. What is the water quality in Ooty lake?
5. What are the cinemas in my city?
6. How many schools are available from 1 km radius of my house and which is the safest route for my children to reach his/her school?
7. What are the overlaps in cell phone service?
8. What is the population within 2 km radius of the nuclear power plant?

GIS Project Design   **8**

8.1 INTRODUCTION

To implement a good GIS project, a well structured project design and management are essential. Well planned project will give fruitful result within the stipulated time. Project not completed in time will not incur more cost and also the time delay will add negative value to the project. Hence all the elements in project planning and implementation should be considered to have a desired result. Some of the issues are also present in GIS and they will be rectified in the next decade.

8.2 PROBLEM IDENTIFICATION

Any project is coming into reality due to necessity. If water scarcity is more in a hard rock region with sufficient rainfall, it is important to augment groundwater resource using rain water harvesting structures. It is important to build water harvesting structures like check dam, percolation pond in suitable locations so that groundwater recharge will be more. Hence a problem is identified and the project work may be carried out by a government or private organisation and the result is given to the government to implement the project. Likewise many area may require a project in GIS environment for planning. Once a project is created any development taking place can be added and GIS database is updated.

8.3 INFRASTRUCTURE REQUIREMENT

A goof infrastructure is required to implement the GIS project. Advanced computers with Pentium 4 processors with high speed and more memory is preferred. Now-a-days multimedia is commonly available in computers. If multimedia component is added, then the GIS project becomes more lively one. A0 colour and monochrome scanners and A0 digitizers are available for data input. A0 colour and monochrome plotters are available for map and image output. All the high end hardware cost more and quality of the data from these hardware are also equally good, but low end hardware give a low quality output. CDs, DVDs and pen drives are available with memory capacity. Data storage is not a problem. Dedicated internet connection and also LAN computers are required to download data and also to share the data. Many GIS Softwares are available in market with varying costs. Before purchasing a software the user should have the capability of different softwares. Experts in GIS may be consulted before purchasing a software. For simple problem, a simple software with low cost is suitable. For advanced analysis high end softwares may be purchased. GIS softwares are coming as modules. A base software will do certain analysis. If one wants to do network analysis, network module of the software may be purchased. Fig 8.1 show the relationship between cost and quality of data. Likewise if a person is interested to do 3D analysis or spatial analysis the respective software may be purchased. Other costs related to rental of a building and related expenses may be considered.

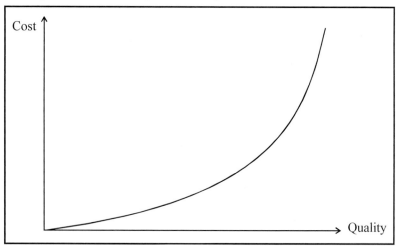

Figure 8.1: Relationship between Cost and Quality

8.4 ORGANIZATION AND GIS EXPERTS

Project is formulated at organization level and head of the organization or head of the GIS project lead the project. He need to monitor the progress of the project periodically. Also sound expertise on GIS is required. Team work is required to implement a project successfully. Digitization of work may be done by experts in digitization and it may be sent to database designers. All the data should be organized and it should be in acceptable form to enter the data in the GIS. Programmers in visual basic is required to customize the GIS. Also integration of multimedia and virtual reality is the trend and experts are required to use these technology with GIS. Cost benefit analysis should be done to reap the benefit out of the GIS project.

8.5 DATA SOURCE

Data is collected in the form of satellite imagery, maps, survey data and attribute data. Project team members should have the knowledge about the imageries available with their characteristics, maps with their characteristics, survey techniques and mode of collection of attribute data. Now-a-days topographic data are available in digital format. Nature of data with their accuracy should be considered. Also different softwares store the data in different formats. During import and export, data is lost. Hence due care should be taken while data is imported or exported from other softwares. Open GIS Consortium has taken this problem, may be rectified in this decade.

8.6 QUESTIONS TO BE ASKED BEFORE PURCHASING A GIS SOFTWARE

1. Is the software useful for my work?
2. Is the software is user friendly
3. Is the software possess most of the analysis module?
4. Will it run in the platform I have?
5. Have any one used the software similar to my work?
6. Can it be upgraded?
7. Do the software vendor give training to me?

8. Is it possible to bring data from other GIS softwares and databases, and CADs?
9. Does it possess macro or allow batch files ?
10. Can it handle large volume of data?
11. Is it possible to convert data from raster to vector or vector to raster?

8.7 Open GIS

This concept was proposed by Open Geospatial Consortium (OGC), a voluntary organization and now most of the GIS software developers are member in OGC. The aim of OGC is to use Data across softwares, across platforms without loss. Presently data is imported from one software to other, then much of the data is lost. This is just like when we bring data from one word processor to other word processor some of the alignments are lost. Similarly during data transfer from one software to other data loss occur. If a software is developed according to OGC standards, then data loss will not occur. So data interoperability is not a problem in near future.

8.8 EVOLUTION IN GIS STANDARD

GIS softwares developed and were advanced during 1990s to 2000. After 2000, most of the GIS analytical capabilities improved and it was the time to develop some standards to share the GIS data across platforms, across computers and across softwares. Hence Some standards were proposed to GIS developers to comply with standards. Later meta data standards were also included. The standalone softwares are not useful because most of the GIS files need to be used by many users. Hence distributed (Shareable) GIS environment came up and later this evolved into Internet GIS (Web GIS). In future data from one GIS software to other software may be read, used and analysed without any editing work.

8.9 FUTURE OF GIS

Computers with memory of 1000 GB(1 terabyte) is going to be available. Touch screen system will be available instead of use of mouse so that even a new GIS user may query a map with his finger. Virtual reality is possible. A person shall virtually go to tour for any area. GIS softwares will use Artificial Intelligence concept to build expert system to do the routine work. Most of the GIS projects have well established methodology. Hence it is possible to train the GIS to do the work automatically and get the result routinely. Fuzzy logic technique will be used to classify the data and extract the data when confusion persists to group the data. Statistical analysis may be further enhanced to find the trend of results.

Colour Plates

Chapter 6: Advanced GIS Applications – Case Studies

Figure 6.6: DEM

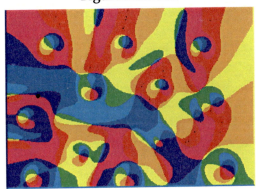

Figure 6.9: Map Aspect

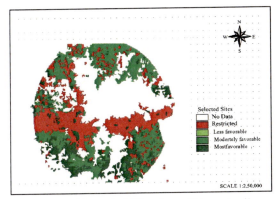

Figure 6.28: Site suitability for solid waste disposal.

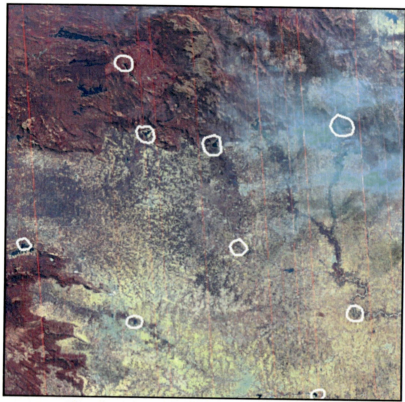

Figure 6.41: Locations of Ground Control Points.

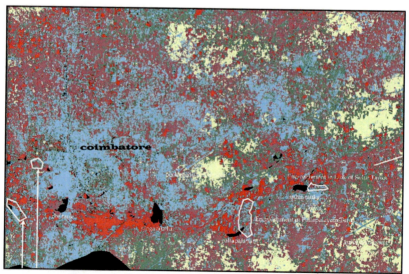

Figure 6.42: Map after supervised classification.

Colour Plates

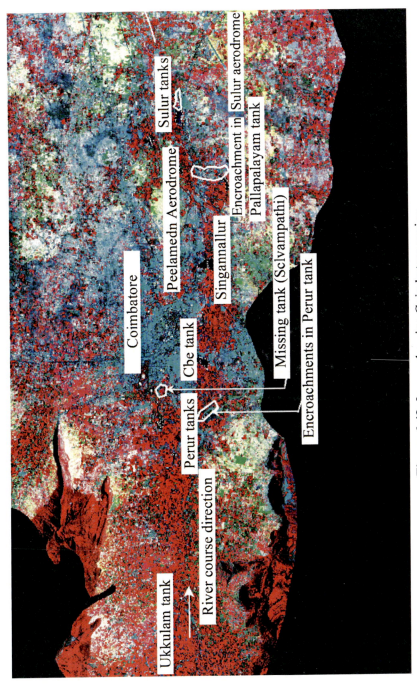

Figure 6.43: Image showing Coimbatore region.

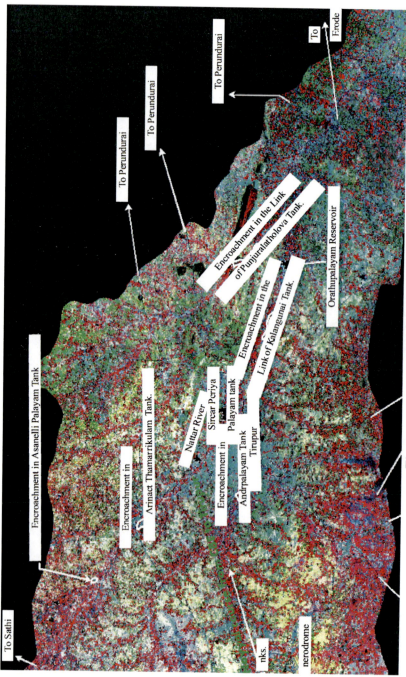

Figure 6.44: Image showing Tirupur region.

Glossary

Absolute location: Points represented in terms of a standard coordinate systems like latitudes and longitudes.

Accuracy: Deviation between the measured value and the true value.

Active X Control: Creation of compound documents consisting of multiple source of information from different applications.

Address matching: A mechanism for relating two files using address as the related item. Geographic coordinates and attributes can be transferred from one address to another. For *e.g.* a data file containing students address can be matched to a street coverage that contains addresses creating a point coverage of where the students live.

Adjacency: A spatial relationship that can be used to select features that share common boundaries.

Affine transformation: Linear transformation (rotation and scaling) and translation (shift of position) used to register maps.

Allocation: A study of spatial distribution of resources through a network.

AM/FM: Automated Mapping and Facility Management (Software for the management of utilities like water, power, telecom lines, wastewater, gas pipelines *etc.*)

Annotation: Text used inside the maps to identify the features.

Arc: A line consisting of a series of vertices joined by straight sections.

ArcInfo Grid: Raster format of ESRI ArcInfo software.

Arithmetic operators: Operators used for multiplication, addition, subtraction of maps.

190   *GIS: Fundamentals, Applications & Implementations*

Artificial Intelligence: A branch of computer science where softwares are used to make the computers think and work. Expert systems are built with artificial intelligence.

Aspect: The directional measure of a slope.

Attribute data: Data that describes characteristics of spatial data.

Automated cartography: Drawing maps in a computer with various drafting packages.

Azimuth: Circle where North is represented as 0° or 360°. East, South, and West directions are represented as 90°, 180° and 270° respecively.

Base map: Background information used to provide context for other layers of information like soil, sampling and location of roads *etc.*

Baud rate: A measure of data transmission speed from computer to other devices. This is like bits per second.

Bench mark: A known point with elevation above a datum established by surveyors for reference and mapping.

Bit: The smallest unit of information that a computer can store and process. A bit has two possible values, 0 or 1 which can be interpreted as yes/no, true/false or on/off.

Block code: Raster data structure which uses square blocks to represent features.

Buffering: A GIS operation in which areas are within a specified distance of selected map features from areas that are beyond.

Byte: A single byte is made up off eight bits *i.e.*

$$\boxed{8 \text{ Bit} = 1 \text{ Byte}}$$

CAD: Computer Aided Design.

Cadastre: "An official register of the ownership, extent, and the value of real property in a given area, used as the basis of taxation" (Random House).

Cartesian coordinate system: A method of representing locations using values along two axis perpendicular to each other.

Cartography: The art of portraying some aspects of the real world as graphical features on a two dimensional surface.

Categorical data: Data measured in terms of description like large, medium and less populated city.

Glossary

Central meridian: A line running North-South at the centre of the graticule

Centroid: Geometric center of a polygon.

CGI: Common Gateway of Interface, CGI scripts allows the owner to obtain information from internet.

Chain code: Raster data structure which stores boundary of features by using cardinal directions and cells.

Check plot: Maps plotted after digitization to check for errors.

Chloropleth map: Map showing homogenous polygons.

Clarke 1866 spheroid: Ground measured spheroid. This is the basis for North American Datum 1927 (NAD 1927).

Closest facility: A network analysis, which combines the shortest paths from the selected location to all candidate facilities and then finds the closest facility among candidates.

Column: A set of cells in database that are vertically represented to store single attributes of every record.

Conformal map: Map which preserves local shapes.

Containment: A spatial relationship that can be used in data query to select features that fall completely within specified features.

Continuous data: Spatial data available between sampled locations.

Contiguous region: A region with joint components.

Contour: Line connecting equal elevations.

Cost distance: Distance measured between location based on cost or friction of existing features.

Coverage: Vector format of ArcInfo.

Dangle: The most common error in GIS digitization, where lines cross each other instead of meeting.

Data compression: Reduction of data volume especially for raster data.

Data exploration: Analysis and query in database.

Datum: Reference plane when measuring elevation.

Database: A computer program for managing an integrated and shared database for data input, search, retrieval, manipulation and output.

192   *GIS: Fundamentals, Applications & Implementations*

Decimal degree system: Longitudes and latitudes are expressed in decimal unit (70.20°, 11.30°).

Delaunary triangle: An algorithm for connecting to form triangles such that all points are connected to their nearest neighbour and triangles are as compact as possible (also called as Voronoi/ Dirichlet triangles).

DEM:. Digital Elevation Model.Computerised representation of an elevation surface. Specifically, a raster image in which the value in each cell represents the surface elevation at that location in the scene.

Destination table: Data brought from the source table to this table.

Source Table --------> Destination Table

DIGEST: Digital Geographic Information Exchange Standard, a format of NATO to exchange maps between member countries.

DIME: Dual Independent Map Encoding, a data standard used by USGS census department.

Digitizer: Input device for vector data. This will store data in x and y coordinates.

Digitizing: A manual method for converting graphical information on paper into digital data.

Discrete data: Geographic features containing boundaries: Point, line or area boundaries.

DLG: Digital Line Graphics Digital representation of USGS quads.

DMS: Degree, Minutes and Seconds System Longitudes and latitudes are represented as DMS (77° 30′ 20″, 11° 02′ 45″). One degree equals to 60 minutes and one minute equals to 60 seconds. /Degree = 60 minutes / Minutes = 60 seconds.

DOD: Digital Orthophoto Quad Digital map derived from satellite imageries and aerial photographs in which the displacement due to camera tilt and terrain relief are rectified.

DPI: Dots per inch, a measure of resolution to print or second

DRG: Digital Raster Graphics, a raster format in which USGS toposheets are stored.

Glossary   193

DTM: Digital Terrain Model. Elevation data in a 3 x 3 arcs seecond grid form or a similar rectilinear form created by the Defense Mapping Agency.

DXF: Data Exchange Format Aformat for storing vector data in ASCII or binary files. Used by AtuoCad and other Cad software for data-interchange. DXF files are convertible into ARC/INFO coverage.

Dynamic segmentation: Vector lines are splited into different types for network analysis.

E00: Data transfer format of ArcInfo.

Easting: A rectangular (x,y) coordinate measurement of distance East from a North-South reference line, usually a meridian used as the axis of origin within a map zone or projection. False porthing is an adjustment constant added to coordinate values to eliminate negative numbers.

Edge matching: Map cleanup function that allows for distortion between adjacent maps to produce a true match of features at the edge of maps. The result is a continuous map, by ensuring that all features that cross the boundary between two adjacent maps appear to be or are in a single theme.

Ellipsoid: Spheroid used to approximate the shape of the earth.

Extent: Coordinates used to define the arieal extent of map in terms of x and y coordinates.

Extrapolation: Estimation of values outside the area of study using the available values in a map.

False northing: An adjustment constant added to northing coordinate values to eliminate negative numbers.

Field: See column

File: Storing related information in computer using unique name.

Filtering: Clarifying detail, sharpening contrast, smoothing edges, and otherwise enhancing image quality flow path. The drainage path through a watershed that begins at any selected point (called the flow path "seed") and runs to one of the outlets of the study site.

Flattening ratio: Ratio between semi-major and semi-minor axis of the earth.

Flow map: Maps created in GIS consists of flow line like drainage with varying thickness.

Font: Text style used in maps like Times New Roman, Arial *etc.*

Format: Data are systematically stored in well established format to be used in many applications and in many platforms.

Freeware: Softwares available for use without any cost.

GCP: Ground Control Points, used in georeferencing an image.

Geocoding: Process of finding locations of a point or attribute in maps.

Geographic coordinates: Method of specifying location in terms of angles measured from the centre of earth (latitudes and longitudes)

Geographic Information System (GIS): A collection of both the hardware and software for dealing with spatial information

Geoid: An irregular equipotential surface closely approximating to global mean sea level.

Georeferencing: The process of using a set of control points to convert images from image coordinates to real world coordinates.

Georelational model: Spatial and attributed data are stored in two different files but are related.

GIS modeling: Use of GIS in building spatially explicit models.

Global Positioning System (GPS) : A method of locating the position of a receiver by measuring the distance from a group of satellites.

Global Positioning Device: Device which uses 24 satellites orbiting at 24200 km above the earth to find latitude, longitude and altitude of any location on the earth.

Graticule: Collection of longitudes and latitudes of the earth.

Greyscale: Image stored in 256 levels of black and white.

Grid cell: See pixel.

Hardware: Devices like computer, scanner, printer, plotter.

Hierarchical database: Data are arranged at different levels and have one to many relationships.

Hypsography: Measurement and mapping of the variations in the earth surface, elevation is in reference to sea level, which are often represented by contour lines on maps.

Glossary 195

Image enhancement: Improving the visual appearance of an image.

Image processing: Converting an useful image into information.

Inter visibility: Places which are visible from a station in a terrain.

Interoperability: Communication between different computer systems. Seamless accessing and sharing of multiple data structures across multiple hardware platforms, operating systems and application software.

Interpolation: Estimation of values at unsampled locations from sampled locations.

Interval data: Data recorded in interval as temperature ($20°$-$30°$ C).

Isoline: Line passing through equal values.

Join: Based on a common field in two tables data are joined from one table (Source table) to other table (Destination table).

Key: A Common field in two tables to link data in relational tables.

Label: Texts placed in maps to identify features in a map.

Land Information System (LIS): Land Information System – Like GIS maps used in conjunction with cadastral or AM/FM maps.

Layer: Creating a single entity in map like a soil map. A layer can also be called as theme.

Legend: A small portion in maps used as a key to identify features in map.

Line: A spatial feature that is represented by a series of points and has the geometric properties of location and length. Also called as chain, arc, link and segment.

Line smoothing: Adding new vertices to make a line smooth for *e.g.* spline.

Location-allocation: Spatial analysis which matches the supply and demand by using sets of objectives and constrains.

Logical operators: AND, OR, NOT used for map overlay (also called as Boolean operators).

Macro: Simple program written in GIS to automate the work.

Map: Abstraction of real world features represented as points, lines and polygons in paper using with cartographic symbolism.

196 *GIS: Fundamentals, Applications & Implementations*

Map projection: The process of transforming from the spherical geographic grid to a plan coordinate system.

Map scale: Ratio of map distance to ground distance.

Matrix: Data with row and column.

Meridians: Imaginary lines running along North-South direction.

Metadata: Information about data.

Model: A simplified representation of phenomenon or system.

Modeling: Simulating real world condition for forecasting by processing data.

Mosaic: Composite of satellite imageries or aerial photographs.

Multispectral: Using more than three bands for capturing images.

Network: A line coverage which is based on the arc-node model and has the appropriate attributes for the flow of objects such as traffic.

Network database: Database with built-in connection across tables.

Node: Ending point of a line or starting point of line.

Nominal data: Data with different categories like soil types.

Normalisation: Reducing non-redundant data.

Northing: Measuring from North of a coordinate system.

Numerical data: Data measured in interval or ratio.

Object oriented database: Spatial and attribute data stored in a single database; well defined relation between objects their properties and behaviour.

Open GIS: Development of common specification and standard in GIS to handle GIS data across various platform and softwares.

Open GIS Consortium (OGC): A non-profit organization to develop Open GIS standards.

Ordinal data: Ranking the data as large medium and small.

Output: Results from GIS are in the form of maps, images, reports, tables and graphs.

Overlay: Arranging several layer one above the other.

Parallels: Imaginary lines running along East-West direction.

Glossary

Photogrammetry: Process of extracting spatial information from aerial photographs.

Pit: Depression in DEM.

Pixel: Picture element, a smallest unit in a raster map.

Point: A spatial feature that is represented by a single coordinate and has only the geometric property of location. It is also called as node or vertex.

Polygon: One or more line (polyline) forming a closed loop.

Precision: Degree of reliability of a measurement.

Proximity: A spatial relationship that can be used to select features that are within a distance of specified features.

Puck: Pointing device on a digitizing tablet.

Quadtree: A method of storing data where the resolution is high *i.e.* where there are a lot of details and low where there are little details.

Raster: Method of storing data in a regular grid or matrix.

Rasterization: Conversion of vector data into raster data.

Ratio Data: Data based on a meaningful zero.

Reclassification: Reclassifying the cell values into more or less than the available classes.

Record: Rows in database to represent a single feature with one or more information. It is also called as Tupule.

Relational database: A database where data in a row or column in a table is related to row and column is other table.

Relative location: Points measured with respect to other locations.

Reprojection: Projecting a map from one projection to other projection.

Resolution: Smallest difference that can be distinguished.

RMSE: Root Mean Square Error: An error between mapped point and actual ground point. This is a positional error arrived during registration of a map.

RLC: Run Length Code: Raster data structure which stores the cell value by row and feature code.

Scanner: Input device which scan paper maps with the help of CCDs and store the data in raster format.

SDTS: Spatial Data Transfer Data Standard : Data formats like USGS, DLG, DEM for transferring data between systems.

Shaded relief map: Map shows the illuminated and shadow region at particular sun's position in sky.

Shortest path analysis: A network analysis which finds path with the minimum cumulative impedance between nodes on a network.

Sliver polygons: An error due to overlay of polygons.

Slope: The rate of change of elevation at a surface location, measured as an angle in degrees or percentage.

Snapping: Process of correcting digitizing errors by making lines or points that should join together.

Source table: Tables from which data are contributed.

Spatial data: Data that describe the geometry of spatial features.

Spline: Polynomial curve used to represent curved features like drainage in maps.

SQL: Structured Query Language adopted in relational databases.

Supervised classification: A type of automatic multi-spectral image interpretation in which the user supervises feature classification by setting up prototypes (collections of sample points) for each feature, class, or land cover to be mapped.

Theme: Set of related geographic features, such as streets, parcels or rivers and their attributes. Geographic features logically organised into groups. Thematic maps emphasizing a single environmental aspect, such as soils, land cover or geology.

TIN: Triangulated Irregular Network : A data model that approximates the terrain with a set of non-overlapping traingles.

Topology: A branch of mathematics applied in GIS to ensure that spatial relationship between features are expressed explicitly.

Traveling salesman problem: A network analysis which finds the best route with the conditions of visiting each stop only once, and returning to the original stop where the journey starts clusters represent.

Glossary

Unsupervised classification: The operation of a group of multi-spectral image interpretation functions (such as K-means) that statistically cluster cells into similar collections.

Utility mapping: Mapping infrastructure systems like water distribution system, sewer lines *etc.*

UTM: Universal Transverse Mercator : One of the most commonly used coordinate system.

Vector: Data structure used to store real world objects in x and y coordinates in computer.

Vectorisation: Conversion of raster data into vector data.

Viewshed: Areas of the land that are visible from an observation point.

Voxel: 3D equivalent of a pixel

Warping: Also called as rubber sheeting to georeference an image using polynomial equation.

Watershed analysis: Creating watershed boundaries drainage network and flow lines.

Workspace: A directory or folder to store related files of a particular project.

WGS84: World Geodetic System 1984 : Satellite determined spheroid.

WYSIWIG: What You See is What You Get. A format which gives matching output of computer screen display and printed image or map.

GIS Resource Banks

Remote Sensing

www.digitalglobe.com
www.spaceimage.com
www.orbimage.com
www.spot.com
www.isro.gov.in
www.nspo.gov.tw
www.nz.dlr.de/moms2p/index.html
www.mentor.eorc.nasda.go.jp/ADEOS
www.alos.nasda.go.jp:80/what.html
www.augusta.co.uk/tentoten
www.earthl.esrin.eas.it
www.ermapper.com
www.pci.on.ca
www.erdas.com
www.rsi.ca

Important GIS web sites

www.esri.com
www.mapinfo.com
www.intergraph.com
www.autodesk.com/map
www.bentley.com
www.idrisi.clarku.edu
www.cecer.army.mil/grass/viz/VIZ.html
www.geo.ed.ac.uk/home/giswww.html
www.itc.nl

Other Sites

www.geog.uu.nl/pcraster.html
www.ncgis.ucsb.edu
www.shef.ac.uk/uni/academic/D-H/gis/gisdata/html
www.usgs.gov/research/gis/title.html
www.altek.com
www.wwwgtco.com
www.laser-scan.com
www.leica-geosystem.com
www.microimages.com
www.trimble.com
www.bluemarble.geo.com
www.Terraserver.microsoft.com
www.infoserver.ciesin.org
www.jpl.nasa.gov
www.gis.com
www.gislinx.com
www.Gisdevelopmnet.net
www.geocomm.com
www.census.gov
www.fws.gov
www.nima.mil
www.nesdis.nooa.gov
www.epa.gov/ngispr
www.girda.no
www.geocan.nrcan.gc.ca
www.auslig.gov.au
www.asprs.org
www.agi.org.uk/pages/links.html
www.geoplace.com
www.spatialnews.com
www.geospatialonline.com
www.urisa.org
www.ced.berkley.edu
www.ucgis.org
www.ogis.org

Magazines and Journals

American Cartographer
Business Geographics
Cartographic Journal
Cartographica
Cartography and Geographic Information Science
Cartography and Geographic Information System
Computers and Geosciences
Computers, Environment and Urban Systems
Geo Asia/Pacific
Geo Europe
Geo Infosystems
Geocarto International
Geodesy, Mapping amd Photogrammetry
Geomatica
Geospatial Today
Geoworld
GISdevelopment.net
GIS India
International Journal of Geographical Information Science
International Journal of Remote Sensing
ISPRS Journal of Photogrammetry and Remote Sensing
Journal of Geographic Information and Decision Analysis (E-Journal)
Journal of Surveying Engineering
Photogrammetric Engineering and Remote Sensing
Photogrammetric Record
Remote Sensing of Environment
Survey Review
Surveying and Land Information System
The ITC Journal

Bibliography

Anderson, J.M., and E.M.Mikhail, 1998: Surveying: Theory and Practice, 2nd ed.McGraw Hill, New York.

Arnoff, S., 1989: Geographic Information Systems: A Management Perspective, Ottawa: WDL Publications.

Atenucci, J.C.(ed.), 1995: Geographic Information Systems: A Guide to Technology, Van Nostrand Reinhold, New York.

Bernhardsen, T., 1999: Geographic Information Systems: An Introduction. 2nd ed. John Wiley & Sons.

Burrough, P.A., 1986: Principles of Geographic Information System for Land Resources Assessment, Oxford University Press, Oxford.

Burrough, P.A., and R.A.McDonnell,1998: Principles of Geographic Information Systems, Oxford University Press, Oxford.

Chrisman. N.R., 2002: Exploring Geographic Information Systems, 2nd ed. John Wiley & Sons.

Clarke, K.C, 1995: Analytical and Computer Cartography, 2nd ed., Englewood Cliffs, NJ: Prentice Hall.

Clarke, K.C., 2001: Getting started with Geographic Information Systems, 3rd ed.Upper Saddle River, NJ:Prentice Hall.

Coppock, J.T and Rhind, D.W., 1991: The History of GIS.

Curran, P., 1985: Principles of Remote Sensing, Longman, London.

D.W.(eds): Geographical Information Systems, Vol.1, Harlow: Longman.

Davis, J.C, 1986: Statistics and Data Analysis in Geology, 2nd ed. John Wiley & Sons.

206   *GIS: Fundamentals, Applications & Implementations*

Demers, M.N, 2000: Fundamentals of Geographic Information Systems, 2nd ed., John Wiley & Sons.

Goodchild, M.F., 1992: Geographical Information Science, *International Journal of Geographic Information Systems*, 6(1): 31-45.

Goodchild, M. and Gopal S. 1989. The Accuracy of Spatial Databases. Taylor & Francis, London.

Guptill, S. and Morrison, J.(eds), 1995: The Elements of Spatial Data Quality. Elsevier, Amsterdam.

Heywood, I.S., Cornelius, and S.Carver, 1998: An Introduction to Geographical Information Systems, Upper Saddle River, Prentice Hall.

Holroyd, F., and Bell, S.B.M., 1992: Raster GIS: Models of Raster Encoding, *Computers and Geosciences*, 18:419-426.

Huxcold, W.E, 1991: An Introduction to Urban Geographic Information Systems, Oxford University Press.

Jensen, J.R., 1996: Introductory Digital Image Processing. A Remote Sensing Perspective, Prentice Hall.

Kang-tsung Chang: Introduction to Geographic Information Systems, 2002, Tata McGraw Hill Publishing Company Ltd., New Delhi.

Kennedy, M., 1996: The Global Positioning System and GIS. Ann Arbor Press Inc.Ann Arbor, 268 pp.

Laurini, R. and D. Thompson, 1992: Fundamentals of Spatial Information Systems, London, Academic Press.

Lillesand, T.M and Kiefer RW, 2000: Remote Sensing and Image Interpretation, 4th ed. John Wiley & Sons, New York.

Lo,C.P., and Yeung K.W., 2002: Concepts and Techniques of Geographic Information Systems, Prentice Hall of India (P) Ltd., New Delhi.

Longley, P.A., M.F.Goodchild, D.J.Maguire, and D.W.Rhind, 2001: Geographic Information Systems and Science, John Wiley & Sons.

Madej, J., 2001: Cartographic Design Ssing ArcView GIS, Albany, Onward Press, New York.

Maguire, D.J., Goodchild,M.F., and Rhind, D.(eds), 1991: Geographical Information Systems: Principles and Applications, Longman Scientific and Technical.

Mailing, D.H., 1992: Coordinate Systems and Map Projections, 2nd ed. Oxford, Pergamon Press.

Bibliography

McHarg. I.L., 1968: Design with Nature, John Wiley and Sons, New York.

Pete Bitinger, and Michael G.Wing, 2004: Geographic Information Systems, Applications in Forestry and Natural Resources Management, McGraw Hill Higher Education, New York.

Peuquet, D.J and D.F.Marble, 1990: Introductory Readings in Geographical Information Systems, Taylor and Francis.

Rhind, D.W., 1992: Data access, charging, and copyright and Their Implications for Geographical Information Systems. International Journal of Geographic Information Systems.

Robinson, A.H., Morrison, J.L., Muehrcke, P.C., Kimerling, A.J and Guptill S.C, 1995: Elements of Cartogarphy, 6th ed., John Wiley & Sons.

Sabins, F., 1997: Remote Sensing : Principles and Interpretation, W.H.Freeman, New York.

Star, J.L., and J.E.Estes, 1990: Geographic Information Systems: An Introduction, Englewood Cliffs, Prentice Hall.

Taylor, J.R., 1982: An Introduction to Error Analysis, University Science Books, Oxford University Press.

Tomlin, C.D, 1990: Geographic Information Systems and Cartographic Modeling, Englewood Cliffs, Prentice Hall.

Tomlinson,R.F., 1984: Geographic Information Systems, The New Frontier, *The Operational Geographer*, 5:31-35.

Wilkinson, G.G, 1996: Review of Current Issues in the Integration of GIS and Remote Sensing Data, *International Journal of Geographic Information Systems*, 10: 85-101.

Wolf, P.R. and C.D.Ghilani, 2002, Elementary Surveying: An Introduction to Geomatics, 19th ed. Prentice Hall, Englewood Cliffs.

Worboys, M.F., 1995: GIS: A Computing Perspective, Taylor & Francis.

Worboys, M.F, Hearnshaw, H.M and Maguire, D.J, 1990: Object Oriented Data Modeling for Spatial Databases, *International Journal of Geographical Information Systems*, 4:369-383.

Worrall, L.(eds.)., 1990: Geographic Information Systems: Developments and Applications, Belhaven Press, London.

Index

A

Accuracy 11, 13, 15, 66, 71, 72, 73, 76, 91, 97, 98, 102, 103, 105, 106, 181, 185
Address Matching 185
Aerial Photograph 66
Affine Transformation 68, 69, 185
Arc 52, 53, 56, 90, 92, 99, 26, 131, 132, 185, 189, 191, 192
Arcinfo 78, 90, 92
Arcview 27, 78, 91, 92, 100, 126, 132, 136, 140, 141, 142, 145, 153, 156, 158, 166, 173, 174
Area 34, 37, 41, 44, 45, 46, 52, 54, 62, 64, 66, 68, 71, 73, 82, 84, 85, 91, 94, 104, 105, 107, 108, 109, 113, 114, 115, 117, 122, 124, 125, 127, 130, 131, 138, 140, 143, 144, 145, 148, 151, 160, 161, 163, 164, 165, 167, 169, 172, 173, 174, 179, 183
Artificial Intelligence 10, 183, 186
Aspatial Data 2, 25, 41, 42
Aspect 18, 31, 34, 51, 72, 115, 119, 120, 123, 186, 194
Attribute Data 2, 3, 12, 24, 25, 26, 29, 34, 52, 77, 78, 92, 131, 145, 176, 181, 186, 192

Autocad 28, 56, 125, 126, 132, 136, 145, 153, 156, 158, 166
Automated Cartography 9, 186
Avhrr 61
Azimuthal Projection 18, 21, 22

B

Bit 55, 56, 90, 109, 186
Block Coding 47, 48
Buffer 35, 54, 113, 136, 145, 146

C

Cad 6, 8, 56, 77, 186,189
Cartography 2, 6, 7, 8, 9, 10, 105, 106, 186
Centroid 187
Chain 48, 49, 113, 187, 191
Chloropleth Map 43, 187
Compression 47, 94, 95, 187
Conical Projection 18, 19, 21
Continuous Data 24, 187
Contour Map 119, 123
Control Point 57
Coordinate System 13, 14, 15, 16, 17, 22, 23, 41, 51, 72, 91, 186, 192, 195
Cost Path 5, 51
Cylindrical Projection 18, 19, 20

D

Data 1, 2, 3, 6, 8, 9, 10, 11, 24, 25, 26, 27, 28, 29, 31, 32, 33, 34, 36, 38, 41, 42, 43, 44, 45, 46, 47, 49, 50, 51, 52, 53, 54, 55, 57, 58, 59, 61, 62, 64, 65, 66, 68, 69, 70, 73, 75, 76, 81, 89, 97, 102, 106, 115, 117, 118, 122, 124, 125, 126, 127, 130, 131, 132, 136, 137, 141, 145, 153, 154, 156, 158, 161, 163, 169, 75, 176, 177, 185, 186, 187, 188, 190, 191, 192, 193, 194, 195

Data Compression 94, 95

Database Management System 3, 8, 77

Datum 6, 14, 17, 72, 75, 98, 186, 187

Delaunary Triangulation 56, 117

Dem 51, 69, 115, 117, 119, 122, 188, 193, 194

DPI 55, 57, 95, 188

Drum Scanner 54

DTM 189

Dynamic Segmentation 111, 189

E

Earth 1, 2, 3, 6, 11, 13, 14, 15, 41, 57, 58, 71, 130, 189, 190

Easting 21, 22, 23, 189

Edge Matching 104

Electromagnetic Spectrum 57, 59, 60

Equator 13, 17, 19, 21

Error 55, 57, 67, 69, 70, 72, 97, 98, 99, 100, 102, 105, 117, 187, 193, 194

Expert System 10, 183

F

Field 1, 2, 6, 28, 34, 37, 38, 44, 62, 70, 74, 75, 77, 78, 83, 84, 97, 124, 126, 130, 131, 132, 153, 165, 189, 191

Flat File 77, 80, 82

Flattening 15, 92, 189, 190

G

Generalization 103, 105

Geocoding 190

Geoid 13, 14, 190

Georeferencing 21, 56, 93, 99, 166, 167, 190

GIS 1, 25, 26, 29, 30, 33, 41, 42, 44, 56, 57, 64, 66, 67, 73, 75, 76, 77, 78, 84, 85, 86, 88, 90, 92, 93, 94, 97, 98, 99, 101, 102, 104, 105, 106, 107, 108, 111, 175, 176, 177, 179, 180, 181, 182, 183, 186, 187, 190, 191, 192, 194

GPS 3, 29, 34, 36, 37, 38, 41, 70, 71, 72, 73, 74, 75, 93, 97, 105, 131, 132, 166, 190

Grid 69

H

Hardware 3, 25, 54, 175, 180, 190, 190, 191

Heads Up Digitization 55, 56

Hierarchical Database Structure 79

I

Image 91, 95, 141, 163, 166, 167

Internet 3, 41, 175, 176, 177, 180, 182

Interpolation 69, 98, 101, 115, 117, 118, 141, 191

Isoline Map 43

L

Landuse Landcover 99
Line 9, 14, 15, 25, 31, 24, 25, 35, 36, 42, 52, 55, 56, 57, 66, 67, 70, 72, 78, 84, 90, 93, 99, 102, 104, 105, 111, 113, 118, 121, 135, 176, 185, 187, 188, 189, 190, 191, 192, 193

M

Map 12, 23, 36, 38, 41, 42, 43, 45, 46, 49, 54, 55, 56, 57, 61, 66, 67, 68, 73, 75, 85, 97, 98, 99, 102, 104, 105, 106, 107, 108, 109, 111, 113, 114, 115, 118, 119, 123, 124, 125, 126, 127, 128, 129, 132, 136, 141, 142, 143, 145, 146, 147,148, 153, 154, 155, 156, 157, 158, 160, 161, 162,163, 165, 166, 167, 168, 175, 176, 177, 180, 183, 186, 187, 188, 189, 190, 191, 192, 193, 194, 195
Metadata 91, 192
Multimedia 10, 11, 107, 180, 181

N

Network 1, 9, 26, 27, 77, 79, 81, 91, 105, 111, 112, 113, 117, 124, 127, 128, 132, 180
Node 52, 53, 70, 99, 102, 112, 113, 192, 193
Northing 21, 22, 23, 189, 192

O

Object Oriented Database 79, 80, 192
Open GIS 11, 28, 92, 182, 192
Overlay 49, 51, 53, 54, 67, 114, 121, 136, 138, 141, 145, 154, 174, 191, 192, 194

P

Pixel 50, 55, 56, 65, 90, 190, 193, 195
Plotter 26, 54, 101, 190
Point 2, 4, 14, 19, 22, 24, 25, 32, 36, 46, 52, 57, 63, 66, 67, 74, 70, 78, 84, 87, 93, 103, 105, 108, 113, 117, 118, 119, 121, 122, 125, 132, 141, 142, 185, 186, 189, 190, 192, 193, 195
Polygon 24, 25, 41, 42, 52, 53, 66, 67, 70, 78, 84, 85, 93, 94, 100, 113, 117, 121, 141, 176
Precision 37, 101, 103, 193

Q

Quadtree 10, 47, 49, 50, 78, 193

R

Raster Data 44, 45, 47, 49, 50, 51, 54
Record 77, 81, 187, 193
Remote Sensing 2, 6, 33, 37, 57, 58, 59, 61, 75, 161
Resolution 11, 12, 29, 45, 46, 49, 51, 53, 55, 57, 61, 62, 63, 91, 95, 98, 161, 163, 188, 193
Rubber Sheeting 69, 195
Run Length Code 95, 193

S

Scale 7, 11, 12, 13, 15, 18, 19, 20, 28, 56, 64, 66, 104, 105, 107, 176
Scanner 54, 55, 190, 194
Simulation 33, 50, 51, 53
Sliver 194
Slope 4, 31, 34, 38, 51, 76, 114, 115, 11, 186, 194

212 GIS: Fundamentals, Applications & Implementations

Snap 70

Software 3, 10, 25, 27, 28, 44, 52, 67, 74, 77, 88, 90, 100, 101, 102, 107, 113, 115, 122, 131, 132, 145, 166, 167, 169, 173, 175, 176, 180, 181, 182, 185, 189, 190, 191

Spaghetti File 51

Spatial Data 2, 3, 6, 11, 25, 36, 41, 42, 77, 78, 87, 91, 97, 125, 186, 187, 194

Spheroid 13, 14, 15, 17, 18, 187, 189, 195

Stream Mode 66, 67

Surveying 2, 33

Surveying 72, 73, 75, 76

Surveying 93, 97

T

Topology 10, 25, 51, 52, 53, 55, 56, 66, 93, 99, 100, 102, 111, 113, 125, 132, 145, 194

Triangulation 56, 57, 71, 72

U

Unsupervised Classification 65

Utilities 8, 34, 36, 37, 75, 106, 124, 130, 185

UTM 20, 21, 166, 167, 195

V

Vector Data Structure 52

Viewshed 195

Voxel 195

W

Weighted Overlay 116, 138

Z

Zone 113, 124, 125, 131, 132, 133, 134, 135, 145, 166, 167

Zone 35, 80, 82, 83, 189